HYDRAULICS

The fundamentals of service and theory of operation for hydraulic systems
in off-road vehicles, trucks and automobiles

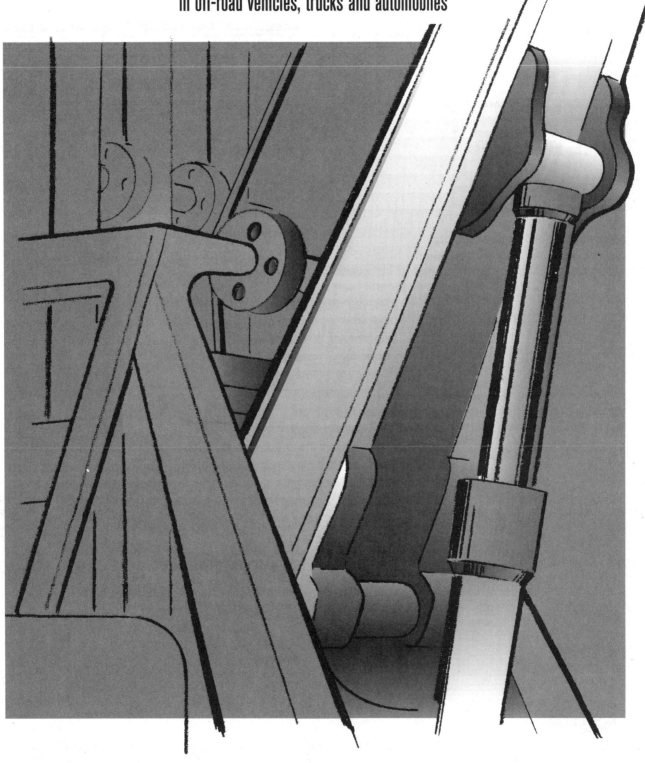

FUNDAMENTALS OF SERVICE

PUBLISHER

Fundamentals of Service (FOS) is a series of manuals created by Deere & Company. Each book in the series is conceived, researched, outlined, edited, and published by Deere & Company. Authors are selected to provide a basic technical manuscript which is edited and rewritten by editors.

PUBLISHER: John Deere Publishing, John Deere Road, Moline, Illinois 61265-8098.

Editor:	John E. Kuhar
Publisher:	Lori J. Lees
Promotions:	Cindy S. Calloway

TO THE READER

PURPOSE OF THIS MANUAL

The main purpose of this manual is to train readers so that they can understand and service hydraulic systems with speed and skill. Starting with "how it works," we build up to "why it fails" and "what to do about it." This manual is also an excellent reference for trained mechanics who want to refresh their memory on hydraulics. It has been written in a simple form using many illustrations so that it can be easily understood.

APPLICATION OF HYDRAULICS IN THIS MANUAL

"Hydraulics" is a broad field. It covers any study of fluids in motion or at rest. But in this manual, the prime interest is in *oil* hydraulics as it is commonly used in agricultural and industrial applications.

HOW TO USE THIS MANUAL

This manual can be used by anyone—experienced mechanics and shop trainees, as well as vocational students and interested laymen.

By starting with the basics, build your knowledge step by step. Chapter 1 covers the basics—"how it works;" Chapters 2 through 10 go into detail about the working parts of hydraulic circuits. Chapters 11 and 12 return to the complete circuit in terms of general maintenance and "why it fails" and how to remedy these failures. Chapter 14 provides important information on Safety.

Persons not familiar with basic hydraulics should start with Chapter 1 and study the chapters in sequence.

Experienced persons can find what they need on the "Contents" page.

Answers to "Test Yourself" questions, at the end of each chapter, appear at the end of this manual on pages 10-12 following the index.

OTHER MANUALS IN THIS SERIES

Other manuals in the FOS series are:

- **Electronic and Electrical Systems**
- **Engines**
- **Power Trains**
- **Shop Tools**
- **Welding**
- **Air Conditioning**
- **Fuels, Lubricants, and Coolants**
- **Tires and Tracks**
- **Belts and Chains**
- **Bearings and Seals**
- **Mowers and Sprayers**
- **Identification of Parts Failures**
- **Glossary of Technical Terms**
- **Fiberglass**
- **Plastics Repair**
- **Fasteners**
- **Hoses, Tubing and Connectors**

Each manual is backed up by a set of 35mm color slides for classroom use. Transparency masters, instructor's guides, and student guides are also available for the first three subjects and this one.

FOR MORE INFORMATION

Write for a free *Catalog of Educational Materials*. Send your request to:

John Deere Publishing
John Deere Road
Moline, IL 61265-8098

**We have
a long-range interest
in good service**

ACKNOWLEDGEMENTS

John Deere gratefully acknowledges help from the following groups: Aeroquip Corporation, American Oil Company, Cessna Aircraft Company, Char-Lynn Company, Commercial Shearing and Stamping Company, Greer Hydraulics, Inc., Gresen Manufacturing Company, Hydreco Division of New York Air Brake Company, National Fluid Power Association, The Nuday Company, Owatonna Tool Company, Sun Oil Company, Sundstrand Corp., Texaco Inc., Vickers Inc.

CONTENTS

Chapter 13 — SYMBOLS USED IN FLUID POWER DIAGRAMS

Chapter 14 — SAFETY RULES FOR HYDRAULICS

DEFINITIONS OF TERMS AND SYMBOLS

METRIC CONVERSION CHART

BOLT TORQUE CHARTS

INDEX

ANSWERS TO TEST YOURSELF QUESTIONS

HYDRAULICS—How it works / CHAPTER 1

BASIC PRINCIPLES OF HYDRAULICS

The basic principles of hydraulics are few and simple:

- **Liquids have no shape of their own.**

- **Liquids are practically incompressible.**

- **Liquids transmit applied pressure in all directions, and act with equal force on all equal areas and at right angles to them.**

- **Liquids provide great increases in work force.**

X 1071

Fig. 1 — Liquids Have No Shape of Their Own.

LIQUIDS HAVE NO SHAPE OF THEIR OWN. They acquire the shape of any container (Fig. 1). Because of this, oil in a hydraulic system will flow in any direction and into a passage of any size or shape.

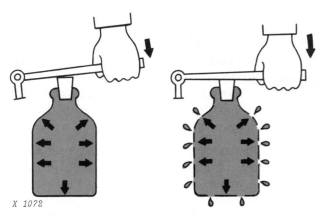

X 1072

Fig. 2 — Liquids Are Practically Incompressible

LIQUIDS ARE PRACTICALLY INCOMPRESSIBLE. This is shown in Fig. 2. For safety reasons, we obviously wouldn't perform the experiment shown. However, if we were to push down on the cork of the tightly sealed jar, the liquid in the jar would not compress. The jar would shatter first. (NOTE: Liquids

will compress slightly under pressure, but for our purposes they are incompressible.)

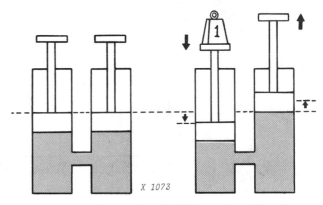

X 1073

Fig. 3—Liquids Transmit Applied Pressure in All Directions

LIQUIDS TRANSMIT APPLIED PRESSURE IN ALL DIRECTIONS. The experiment in Fig. 2 shattered the glass jar and also showed how liquids transmit pressure—in all directions when they are put under compression. This is very important in a hydraulic system. In Fig. 3, take two cylinders of the same size (one square inch) and connect them by a tube. Fill the cylinders with oil to the level shown. Place in each cylinder a piston which rests on the columns of oil. Now press down on one cylinder with a force of one pound. This pressure is created throughout the system, and an equal force of one pound is applied to the other piston, raising it as shown.

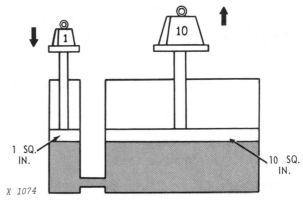

1 SQ. IN.

10 SQ. IN.

X 1074

Fig. 4—Liquids Provide Great Increases in Work Force

LIQUIDS PROVIDE GREAT INCREASES IN WORK FORCE. Now let's take two more cylinders of different sizes and connect them as shown in Fig. 4. The first cylinder has an area of one square inch, but the second has an area of ten square inches. Again use a force of one pound on the piston in the smaller cylinder. Once again the pressure is

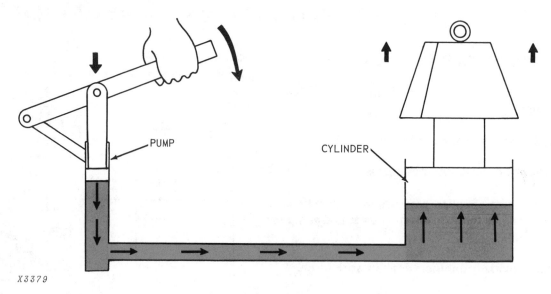

Fig. 5—A Basic Hydraulic System

created throughout the system. So a pressure of one pound per square inch is exerted on the larger cylinder. Since that cylinder has a piston area of ten square inches, the total force exerted on it is *ten* pounds. In other words, we have a great increase in work force.

This principle helps you to stop a large machine by pressing a brake pedal.

The force (F) exerted by a piston can be determined by multiplying the piston area (A) by the pressure (P) applied.

 CAUTION: The forces in a hydraulic system can be very high. Use extreme caution. See Chapter 14 for safety instructions.

HOW A HYDRAULIC SYSTEM WORKS

Let's build up a hydraulic system, piece by piece.

The basic hydraulic system has two parts:

1. The PUMP which moves the oil.

2. The CYLINDER which uses the moving oil to do work.

In Fig. 5, when you apply force to the lever, the hand *pump* forces oil into the *cylinder*. The pressure of this oil pushes up on the piston and lifts the weight.

In effect, the *pump* converts a mechanical force to hydraulic power, while the *cylinder* converts the hydraulic power back to mechanical force to do work.

But for continued operation of the system, we must add some new features (Fig. 6).

3. CHECK VALVES to hold the oil in the cylinders between strokes and to prevent oil from returning to the reservoir during the pressure stroke. The ball-type valves open when oil is flowing but close when the flow stops.

4. A RESERVOIR to store the oil. If you keep on pumping to raise the weight, a supply of extra oil is needed. The reservoir has an air vent which allows oil to be forced into the pump by gravity and atmospheric pressure when the pump piston is retracted.

Notice that the pump is smaller than the cylinder. This means that each stroke of the pump would only move enough oil to move the piston a small amount. However, the load lifted by the cylinder is much greater than the force applied to the pump piston. If you want to lift the weight faster, then you must work the pump faster, increasing the volume of oil to the cylinder.

The system we have just described is a system which might be found on a hydraulic jack or a hydraulic press; however, to meet the hydraulic requirements in most other applications, we must provide a greater quantity of oil at a more consistent rate and also have better control of the oil movement.

Let's complete the circuit and add some new features as shown in Figs. 7 and 8.

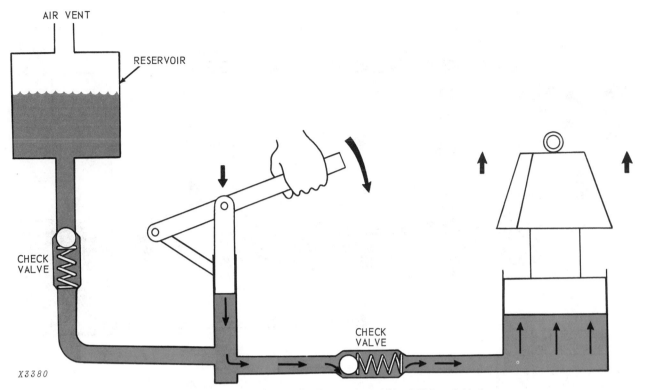

Fig. 6—Hydraulic System with Reservoir and Check Valves Added

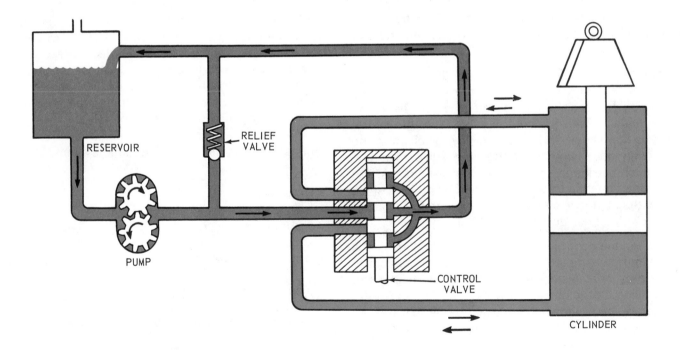

Fig. 7 — Hydraulic System with Relief Valve and Double-Acting Cylinder Added

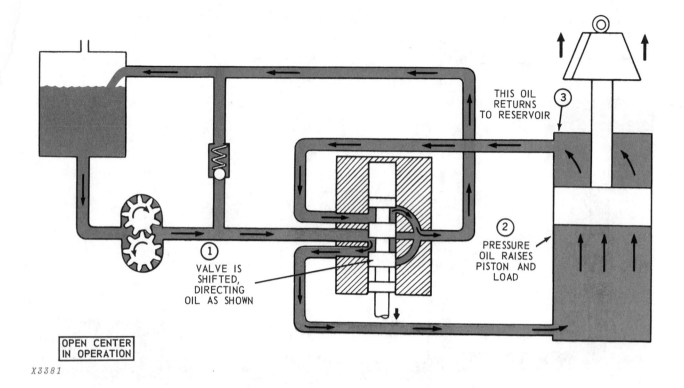

OPEN CENTER
IN OPERATION

X3381

Fig. 8—Hydraulic System in Operation—Raising a Load (Open-Center Type)

We have now added a *gear-type pump.* This is one of many types of pumps which transform the rotary force of a motor or engine to hydraulic energy. For more on pumps, see Chapter 2.

5. The CONTROL VALVE directs the oil. This allows the operator to control the constant supply of oil from the pump to and from the hydraulic cylinder. When the control valve is in the neutral position shown in Fig. 7, the flow of oil from the pump goes directly through the valve to a line which carries the oil back to the reservoir. At the same time, the valve has trapped oil on both sides of the hydraulic cylinder, thus preventing its movement in either direction.

When the control valve is moved down (Fig. 8), the pump oil is directed to the cavity on the bottom of the cylinder piston, pushing up on the piston and raising the weight. At the same time, the line at the top of the cylinder is connected to the return passage, thus allowing the oil forced from the top side of the piston to be returned to reservoir.

When the control valve is moved up (not shown), oil is directed to the top of the cylinder, lowering the piston and the weight. Oil from the bottom of the cylinder is returned to the reservoir.

6. The RELIEF VALVE protects the system from high pressures. If the pressure required to lift the load is too high, this valve opens and relieves the pressure by dumping the oil back to the reservoir. The relief valve is also required when the piston reaches the end of the stroke. At this time there is no other path for the oil and it must be returned to the reservoir through the relief valve.

This completes our basic hydraulic system.

SUMMARY

To summarize:

• **The pump = generating force**

• **The cylinder = working force**

• **The valve = oil flow and direction control**

• **The reservoir = oil storage**

For more details on how these parts operate, refer to Chapter 2—Pumps, Chapter 3—Valves, Chapter 4—Cylinders, and Chapter 8—Reservoirs.

THE PROS AND CONS OF HYDRAULICS

As you have seen in the simple hydraulic system, we have just developed, the purpose is to transmit power from a source (engine or motor) to the location where this power is required for work.

To look at the advantages and disadvantages of the hydraulic system, let's compare it to the other common methods of transferring this power. These would be mechanical (shafts, gears, belts, chains, or cables) or electrical.

ADVANTAGES

1. FLEXIBILITY—Unlike the mechanical method of power transmission where the relative positions of the engine and work site must remain relatively constant with the flexibility of hydraulic lines, power can be moved to almost any location.

2. MULTIPLICATION OF FORCE—In the hydraulic system, very small forces can be used to move very large loads simply by changing cylinder sizes.

3. SIMPLICITY—The hydraulic system has fewer moving parts, fewer points of wear. And it lubricates itself.

4. COMPACTNESS—Compare the size of a small hydraulic motor with an electric motor of equal horsepower. Then imagine the size of the gears and shafts which would be required to create the forces which can be attained in a small hydraulic press. The hydraulic system can handle more horsepower for its size than either of the other systems.

5. ECONOMY—This is the natural result of the simplicity and compactness which require relatively low cost for the power transmitted. Also, power and frictional losses are comparatively small.

6. SAFETY—There are fewer moving parts such as gears, chains, belt and electrical contacts than in other systems. Overloads can be more easily controlled by using relief valves than is possible with the overload devices on the other systems.

DISADVANTAGES

1. EFFICIENCY—While the efficiency of the hydraulic system is much better than the electrical system , it is lower than for the mechanical transmission of power.

2. NEED FOR CLEANLINESS—Hydraulic systems can be damaged by rust, corrosion, dirt, heat and breakdown of fluids. Cleanliness and proper maintenance are more critical in the hydraulic system than in the other methods of transmission.

COMPARING HYDRAULIC SYSTEMS

Two major types of hydraulic systems are used today:

● **Open-Center Systems**

● **Closed-Center Systems**

The simple hydraulic system which we developed earlier in this chapter (Fig. 8) is what we call an OPEN-CENTER SYSTEM. This system requires that the control valve spool be *open* in the center to allow pump flow to pass through the valve and return to the reservoir. The pump we have used supplies a *constant* flow of oil and the oil must have a path for return when it is not required to operate a function.

In the CLOSED-CENTER SYSTEM, the pump is capable of "taking a break" when oil is not required to operate a function. Therefore, the control valve is *closed* in the center, which stops (dead ends) the flow of oil from the pump—the "closed center" feature.

The open-center system is shown in neutral position in Fig. 9 on the next page, while the closed-center system is shown in Fig. 10.

To summarize:

● **Open-Center System—oil is pumped constantly, with valve open in center to allow oil to return to reservoir.**

● **Closed-Center System — valve spool closed in center to dead end pump oil in neutral.**

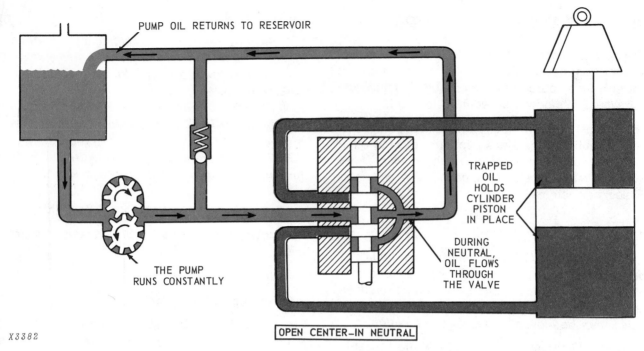

Fig. 9—Open-Center System in Neutral

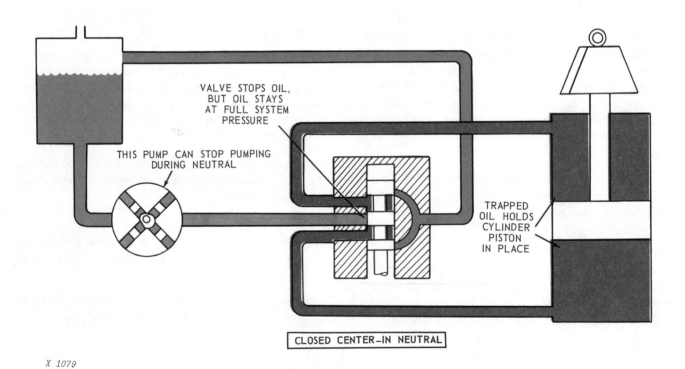

Fig. 10—Closed-Center System in Neutral

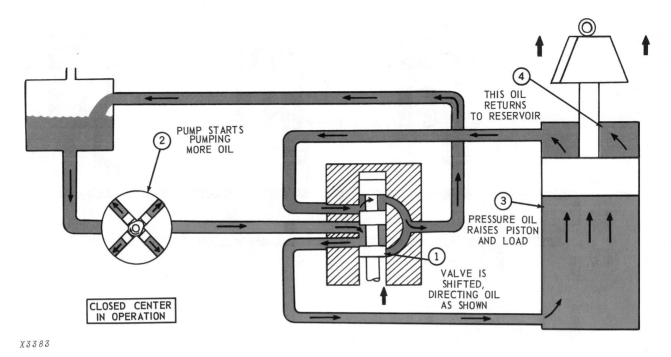

X3383

Fig. 11—Closed-Center System in Operation—Raising a Load

CLOSED CENTER SYSTEM

Let's look at a closed-center system with a variable displacement pump.

In neutral, the pump pumps oil until pressure rises to a predetermined level. Then a pressure regulating valve allows the pump to shut itself off and to maintain this pressure to the valve.

When the control valve is operated as shown in Fig. 11, oil is diverted from the pump to the bottom of the cylinder.

The drop in pressure caused by connecting the pump pressure line to the bottom of the cylinder causes the pump to go back to work, pumping oil to the bottom of the piston and raising the load.

When the valve was moved, the top of the piston was connected to a return line, thus allowing return oil forced from the piston to be returned to the reservoir or pump.

When the valve is returned to neutral, oil is again trapped on both sides of the cylinder and the pressure passage from the pump is dead ended. At this time the pump again takes a break.

Moving the spool in the downward position (not shown), directs oil to the top of the piston, moving the load downward. Then oil from the bottom of the piston is sent into the return line.

With the closed center system, if the load exceeds the predetermined standby pressure or if the piston reaches the end of its stroke, the pressure build-up simply tells the pump to take a break, thus eliminating the need for relief valves to protect the system.

We have now built the simplest of open- and closed-center systems. However, most hydraulic systems require their pump to operate more than one function.

Let's look at how this is done and compare the advantages and disadvantages of each system.

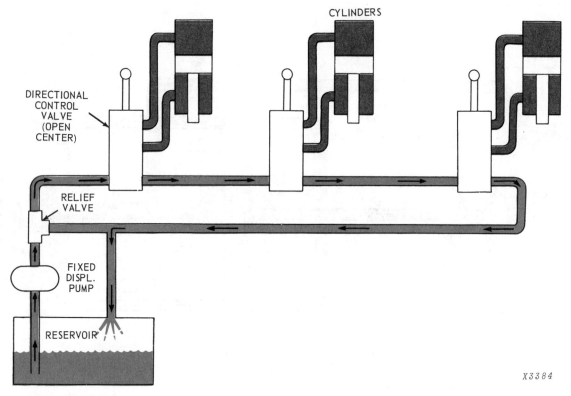

Fig. 12—Open-Center System with Series Connection

VARIATIONS ON OPEN- AND CLOSED-CENTER SYSTEMS

To operate several functions at once, hydraulic systems have the following connections:

OPEN-CENTER SYSTEMS

- *Open-Center with Series Connection*
- *Open-Center with Series Parallel Connection*
- *Open-Center with Flow Divider*

CLOSED-CENTER SYSTEMS

- *Closed-Center with Fixed Displacement Pump and Accumulator*
- *Closed-Center with Variable Displacement Pump*

Let's discuss each of these systems.

OPEN-CENTER SYSTEMS

Open-Center System with Series Connection

Fig. 12 shows a series connection of the open-center system. Oil from the pump is routed to the three control valves in series. The return from the first valve is routed to the inlet of the second, etc.

In neutral, the oil passes through the valves in series and returns to the reservoir as shown by the arrows. When a control valve is operated, incoming oil is diverted to the cylinder which that valve serves.

Return oil from the cylinder is directed through the return line and on to the next valve.

This system is satisfactory as long as only one valve is operated at a time. In this case the full output of the pump at full system pressure is available to that function. However, if more than one valve is operated, the total of the pressures required for each individual function cannot exceed the system relief setting.

Open-Center System with Series Parallel Connection

This system, shown in Fig. 13, is a variation on the series connected type. Oil from the pump is routed through the control valves in series—but also in parallel. The valves are sometimes "stacked" to allow for the extra passages.

In neutral, the oil passes through the valves in series as shown by the arrows. But when any valve is operated, the return is closed and oil is available to *all* valves through the parallel connection (upper blue line).

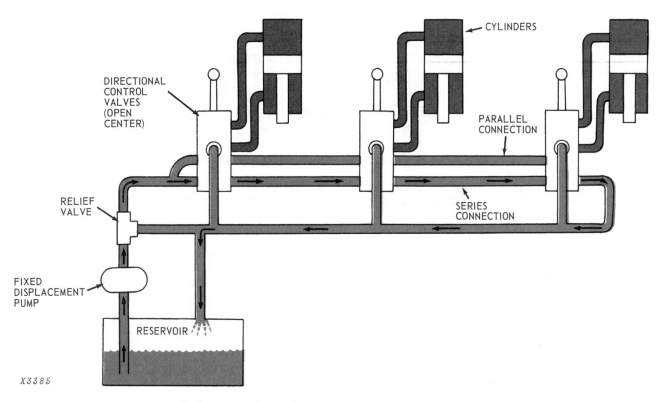

X3385

Fig. 13—Open-Center System with Series Parallel Connection

When two or more valves are operated at once, the cylinder which needs least pressure will operate first, then the next least, etc. However, this ability to satisfy two or more functions at once is an advantage over the *series* connection in Fig. 12.

Open-Center System with Flow Divider

Fig. 14 shows a flow divider used with an open-center system. The flow divider takes the volume of oil from the pump and divides it between two functions. For example, the flow divider might be designed to open the left side first in case both control valves were actuated at the same time. Or it might divide oil to both sides—either equally or by percentage. With this system, the pump must be large enough to operate all the functions at once. And the pump must supply all this oil at the maximum pressure of the highest function. This means that a lot of horsepower is wasted when operating only one control valve.

We can see now that while the open-center system is efficient on single functions, it has a limited value for use with multiple functions.

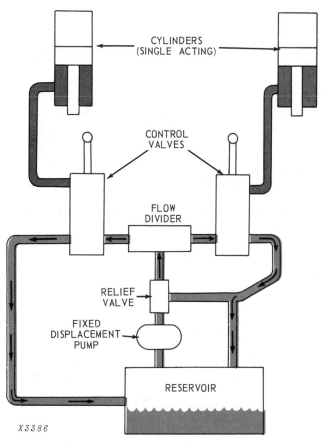

X3386

Fig. 14—Open-Center System with Flow Divider

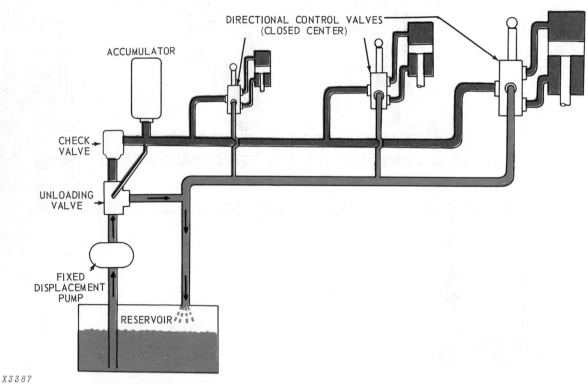

DIRECTIONAL CONTROL VALVES
(CLOSED CENTER)

ACCUMULATOR

CHECK
VALVE

UNLOADING
VALVE

FIXED
DISPLACEMENT
PUMP

RESERVOIR

X3387

Fig. 15—Closed-Center System with Fixed Displacement Pump and Accumulator

CLOSED-CENTER SYSTEMS

Closed-Center System with Fixed Displacement Pump and Accumulator

This system is shown in Fig. 15. A pump of small but constant volume charges an accumulator. When the accumulator is charged to full pressure, the unloading valve diverts the pump flow back to the reservoir. The check valve traps pressure oil in the circuit.

When a control valve is operated, the accumulator discharges its oil and actuates the cylinder. As pressure begins to drop, pump flow is again directed by the unloading valve to the accumulator to recharge it.

This system, using a small capacity pump, is effective when operating oil is needed only for a short time. However, when the functions need a lot of oil for longer periods, the accumulator system cannot handle it unless the accumulator is very large.

Closed-Center System with Variable Displacement Pump

This system is shown in Fig. 16. We have already shown much of this system in Fig. 10, but now we are adding a charging pump. This pumps oil from

the reservoir to the variable displacement pump. The charging pump supplies only the make-up oil required in the system and provides some inlet pressure to make the variable displacement pump more efficient. Return oil from the system functions is sent directly to the inlet of the variable displacement pump as shown.

We saw earlier that the open-center system is the simplest and least expensive for hydraulic systems which have only a few functions. But as more functions are added with varying demands for each function, the open-center system requires the use of flow dividers to proportion the oil flow to these functions. The use of these flow dividers in an open-center system reduces efficiency with resulting heat build-up.

Today's machines need more hydraulic power and *the trend has been to the closed-center system.*

On a modern tractor, for example, oil may be required for power steering, power brakes, remote cylinders, three-point hitch, loaders, and other mounted equipment.

In most cases, each of these functions has a requirement for different quantities of oil. With the closed-center system, the *quantity of oil to each function can be controlled* by line size, valve size, or by orificing with less heat build-up when com-

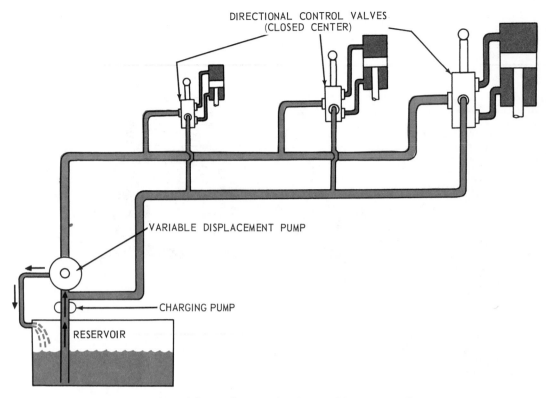

Fig. 16—Closed-Center System with Variable Displacement Pump

pared to the flow dividers necessary in a comparable open-center system.

Other Advantages of Closed-Center Systems

1. There is *no requirement for relief valves* in a basic closed-center system because the pump simply shuts itself off when standby pressure is reached. This prevents heat build-up in systems where relief pressure is frequently reached.

2. The *size of lines, valves, and cylinders can be tailored* to the flow requirements of each function.

3. By using a larger pump, *reserve flow is available* to insure full hydraulic speed at low engine rpm. More functions can also be served.

4. On functions such as brakes which require force but very little movement on a piston, the closed-center system is very efficient. By holding the valve open, *standby pressure is constantly applied* to the brake piston with no loss of efficiency because the pump has returned to standby.

In a similar open-center system, the pump would operate in relief to maintain this pressure.

THE USES OF HYDRAULICS

Hydraulic power can be adapted to thousands of uses. A few of the major uses on the farm and in industry will be covered here.

Hydraulics can be used at several points on a single machine. The tractor in Fig. 17 uses hydraulics to steer, brake, control mounted equipment, and supply remote operation of tools. A single hydraulic system serves to power all these functions.

Let's briefly discuss the major uses of hydraulics.

HYDRAULIC STEERING SYSTEMS

Three major types of steering are used for today's machines:

1. MANUAL STEERING

2. POWER STEERING
    a. Hydraulic steering with mechanical drag link
    b. Hydrostatic steering
    c. Metering pump steering

3. HYDRAULIC ASSIST STEERING

Let's look at each type.

Manual Steering

The steering wheel is linked directly to the turning wheels and the operator does all the work of steering. No hydraulics is used—only mechanical effort.

Power Steering

With full power steering, the only force required from the operator is enough steering wheel force

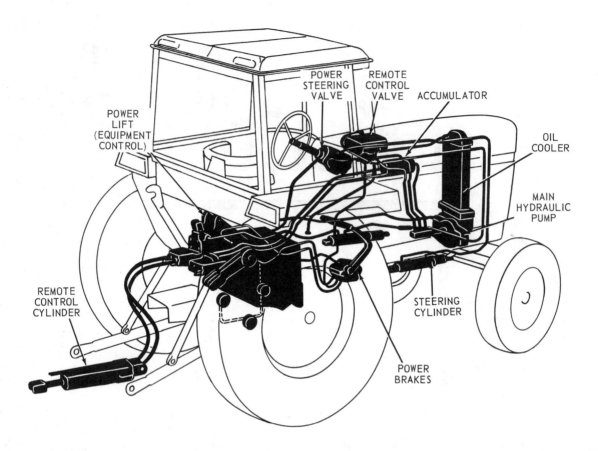

POWER
STEERING
VALVE

REMOTE
CONTROL
VALVE

ACCUMULATOR

POWER
LIFT
(EQUIPMENT
CONTROL)

OIL
COOLER

MAIN
HYDRAULIC
PUMP

REMOTE
CONTROL
CYLINDER

STEERING
CYLINDER

POWER
BRAKES

Fig. 17—A Modern Tractor With Full Hydraulics (Closed-Center System Shown)

to open the valves. Hydraulic power is supplied by a pump which gives all steering force up to the capability of the system.

These systems are divided into three major categories.

A. HYDRAULIC STEERING WITH MECHANICAL DRAG LINK

Fig. 18 illustrates hydraulic steering with a mechanical drag link. We are showing it with the open-center hydraulic system; however, it is equally adaptable to the closed-center system. Operation is shown during a right turn.

In the right turn, the operator turns the steering wheel to the right as shown. Because of the resistance in turning the steering wheels, the shaft is forced up out of the worm nut. This shifts the spool valve and the steering shaft up, which directs oil to the cylinder at the front wheels. This cylinder rotates a rack and pinion device which turns the front wheels. Oil from the other side of the steering cylinder is returned through the spool valve to the reservoir as shown.

As long as the steering wheel is turned, oil will continue to move the wheels. As soon as the steer-

ing wheel motion is stopped, the hydraulic pressure will turn the wheels slightly further to the right, moving the steering linkage forward and pulling the valve back to the neutral position.

B. HYDROSTATIC STEERING

Hydrostatic steering has no mechanical connection between the steering valve and the steering cylinders. Basically the operation is the same as that just described except that we have a hydraulic "drag link" instead of a mechanical one.

Fig. 19 on the next page shows a hydrostatic steering system used with the closed-center hydraulic system. Operation is shown during a right turn.

When the operator turns the steering wheel to the right, the steering shaft, which is threaded through the steering valve piston, attempts to pull this piston upward. Because oil is trapped in the circuit at this time, the shaft instead moves the collar downward, rotating the pivot lever and opening a pressure and return valve. When the valves open, pressure oil enters the steering valve cylinder, forcing the piston upward. This pushes the oil out of the valve cylinder and into the right-hand steering cylinder, turning the front wheels to the right.

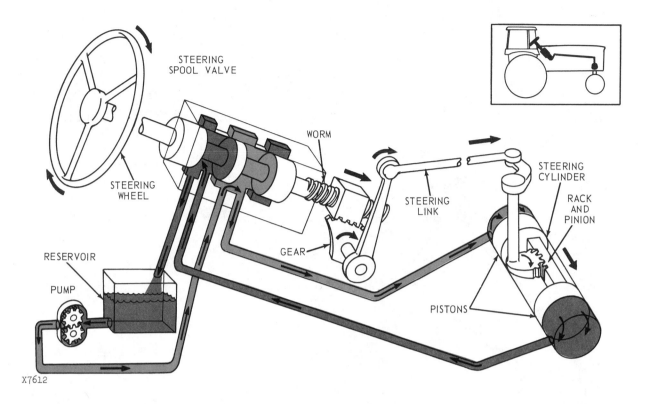

Fig. 18—Hydraulic Steering During A Right Turn

As the wheels turn, oil is forced out of the left-hand steering cylinder and returns through the open return valve to the reservoir or pump.

When the operator stops turning the steering wheel, the steering shaft is moved upward by the steering valve cylinder, pulling the collar upward and centering the pivot lever, thus closing the valves.

During the left turn, the operation is reversed. (For simplicity, the valves used during a left turn are not shown.)

C. METERING PUMP POWER STEERING

Metering pump power steering consists of three assemblies (Fig. 20):

- **metering pump**

- **steering valve**

- **steering motor**

As with hydrostatic steering, there are no mechanical connections between assemblies in metering pump steering.

Note: Indications of directions refer to those as seen from the operator's seat. Side A and side B will help identify direction of movement in the steering valve housing.

Fig. 20 shows the metering pump power steering operating during a right turn. When the operator turns the wheel to the right the gears in the metering pump direct oil in the steering system to the steering valve housing and to the left end of the feedback piston. This oil (under some pressure) moves the steering valve toward side B.

The movement of the steering valve opens the pressure oil circuit to the left end of the steering piston. Oil from the right end of the piston flows back to the steering valve oil gallery and to the reservoir.

Oil from the right end of the feedback piston cylinder is forced out, by piston movement, and returns through the steering valve housing to the metering pump.

This movement of the steering and feedback pistons from left to right causes the spindle to rotate clockwise and turn the front wheels to the right.

When the operator stops turning the steering wheel, the gears in the metering pump stop directing oil to the steering valve. Circuit pressure, from the right end of the feedback piston to the steering valve, (caused by movement of the feedback piston) acts against the side B of the steering valve. The valve moves toward side A, closing the pressure oil

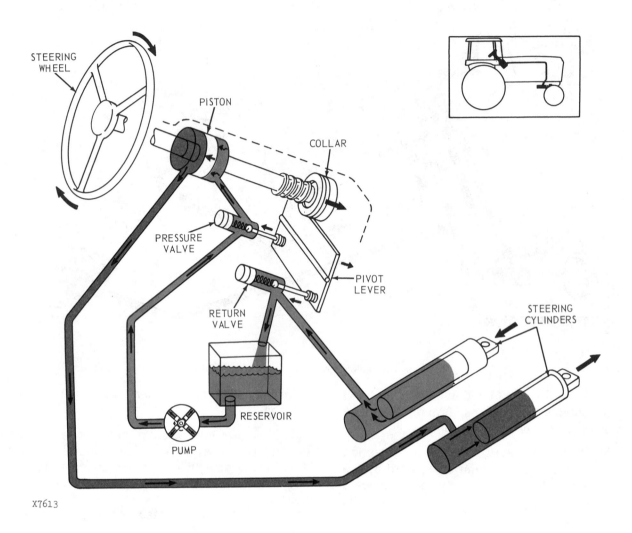

Fig. 19—Hydrostatic Steering During a Right Turn

passage from the main hydraulic pump, and stops the turning movement. The valve becomes centered and traps oil in passages to both sides of the steering piston. The trapped oil holds the wheels in the right turn until the operator again turns the steering wheel.

If oil is lost from the control circuit, pressure in the control circuit drops. The reduced pressure causes the make-up valve to unseat and allow return oil from the steering piston to make up oil in the control circuit.

On articulated tractors, the steering and feedback pistons are replaced by hydraulic cylinders which control steering.

Manual Turn with Metering Pump Steering

When there is no inlet pressure oil to the steering valve housing, the machine may be steered manually. Without pressure oil, the inlet check valve is seated preventing oil in the steering system from entering the hydraulic system pressure circuit.

When the operator turns the wheel to the right (Fig. 21) oil in the steering system is again directed to side A of the steering valve and the left end of the feedback piston. Enough pressure is exerted on side A of the steering valve to unseat the manual steering check valve in the hollow steering valve.

Oil then passes through the steering valve to the left end of the steering piston. The force of the oil on the feedback piston and steering piston moves both pistons to the right, turning the steering spindle clockwise, and turning the front wheels to the right.

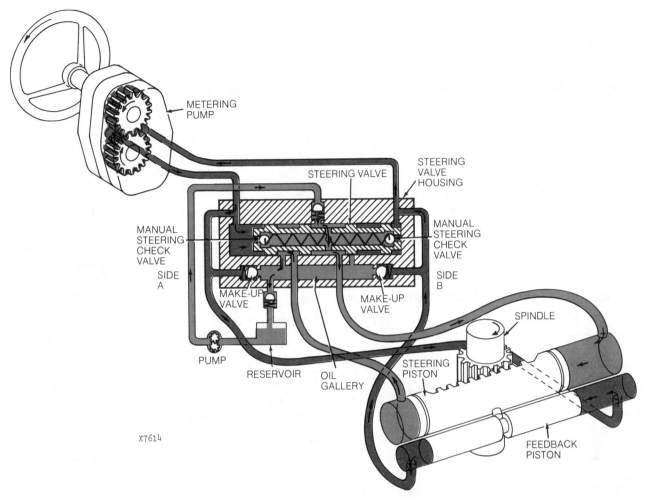

Fig. 20 — Metering Pump Steering During a Right Turn

Oil from the right end of the feedback piston cylinder is forced out, by piston movement, and returns through the steering valve housing to the metering pump.

Oil from the right end of the steering piston is forced to return to the oil gallery. But this oil does not return to the reservoir. Instead, it opens the make-up valve

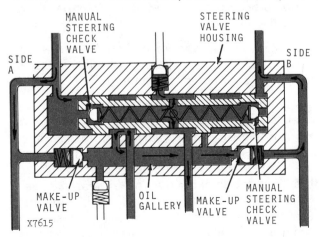

on side B and joins with oil returning from the feedback piston cylinder. This insures a recirculating oil supply in the steering circuit.

When the operator stops turning the steering wheel, the gears in the metering pump stop directing oil to the steering valve.

The manual steering check valve seats and stops the turning movement. All oil in the steering system becomes trapped and holds the wheels in a right turn until the operator again turns the steering wheel.

Hydraulic Assist Steering

In hydraulic assist steering systems (not shown), steering force is supplied by a combination of manual effort and hydraulic power. It is used in systems where the operator must maintain a high degree of feel in the steering.

In these systems, the amount of pressure built up in the steering system is in direct proportion to the amount of effort used by the operator. The most common use of this system is in the power steering assist for crawler tractors.

Protection Against Failure of Power Steering

If oil steering power is lost in hydraulic steering with a mechanical drag link, the solid link takes over and the operator can still steer the machine mechanically, but with more effort.

The same protection is also provided in the hydro-static power steering. This is done by trapping oil in the steering valve and using the steering valve piston as a motor. In a right turn, the steering valve piston is pulled upward, forcing oil into the right-hand steering cylinder. In a left turn, the steering valve piston is forced down, pushing oil into the left-hand piston.

HYDRAULIC AND POWER BRAKE SYSTEMS

Three major types of brakes are used to turn or stop farm and industrial machines:

- **Manual brakes**

- **Hydraulic brakes**

- **Power brakes**

1. MANUAL BRAKES. When the operator applies the brakes, a mechanical link causes dry brake disks, or shoes to slow the rolling wheels by friction.

2. HYDRAULIC BRAKES. When the operator applies the brakes, he pushes a column of trapped oil which clamps the brake disk or shoes to slow the rolling wheels.

3. POWER BRAKES. When the brakes are applied, fluid power takes over and slows the wheels.

On some machines, two types of brakes may be used. For example, the power brakes for stopping may be backed up by a manual brake for parking. Both brakes may actuate the same braking mechanism.

On most tractors, brakes are located on each rear wheel. For turning, the operator presses down the pedal for the left or right wheel. For stopping, he presses down on both pedals at once.

On four-wheel drive machines, a single brake mechanism at the transmission usually controls the whole unit.

Let's discuss the operation of hydraulic and power brakes in more detail.

Hydraulic Brakes

Fig. 22 on the next page shows hydraulic brakes on a typical tractor. Operation is shown during a left turn.

For a sharp left turn, the operator presses down the left brake pedal. This rotates the pedal arm against the brake piston as shown and moves it to the rear. The piston closes the inlet check valve from the reservoir, trapping oil in the cylinder. As the piston moves farther, it forces the trapped oil out of the cylinder, unseating the outlet check valve.

The oil is pushed through a pipe to the final drive at the left rear axle, where it applies force against the brake pressure plate (see inset). This presses the revolving brake disk against the side of a fixed plate, braking the left axle and wheel.

When the brake pedal is released, the force against the brake disk is relieved. Oil returning from the axle unit is metered by the outlet check valve and retainer (inset, Fig. 22). Spring pressure pushes the piston to the front again. This creates a vacuum in the cylinder, allowing the reservoir check valve to open again. More oil then enters the cylinder as needed for the next braking.

When both brake pedals are pressed down at once, oil is sent by both brake valves to both final drives. To assure equal oil pressure on both sides, equalizing valves under each brake piston are opened, connecting the two brake cylinders.

If the tractor is stopped or the oil supply fails, braking is still possible using the oil in the brake reservoirs. Normally, oil is supplied to the brake reservoirs from the tractor hydraulic system.

Power Brakes

"Power" brakes mean that hydraulic force completely controls the braking of the machine, once the operator presses down the brake pedal to actuate the valving.

Fig. 23 shows full power brakes on a modern wheel tractor with a closed-center hydraulic system. Operation is shown during a left turn.

For a sharp left turn, the operator presses down on the left brake pedal. This causes a rod linkage to push down on the brake valve and open it. Inlet oil under pressure now rushes in through the open valve, forces the outlet check valve open, and flows on to the final drive at the left rear axle (see inset). Here the oil forces the brake pistons and pressure plates to press the revolving brake disk against a fixed plate and so brakes the left axle and wheel.

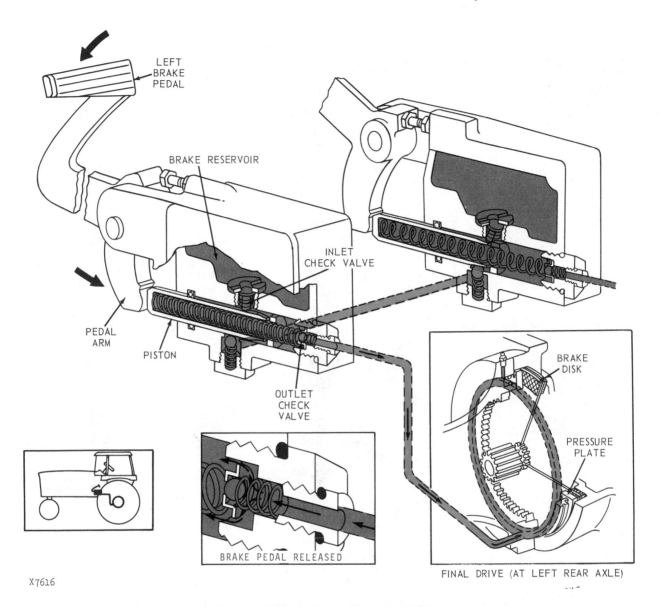

Fig. 22 — Hydraulic Brakes During a Left Turn

When the brake pedal is released, the brake valve is closed again by its spring, and inlet oil is shut off. This relieves pressure on the brake disk at the axle, and braking stops as some oil flows back to the brake valve area. This oil dumps into the brake reservoir after going past the valve and through the valve plunger.

When both brake pedals are pressed down at once, oil is sent by both brake valves to both final drives. To assure equal braking, equalizing valves (not shown) are opened, connecting the two brake valves.

PROTECTION AGAINST FAILURE OF POWER BRAKES

If the tractor is stopped or the pressure oil supply fails, braking is still possible using the oil in the brake reservoir. The "power" brakes then become "hydraulic" brakes, using trapped oil to brake the tractor.

Operation is as follows: As pressure fails, the inlet check valve closes and oil is trapped in the brake valve area. Then if the left brake pedal is pushed down, the brake valve bore becomes a cylinder and piston device, forcing oil out to the rear axle unit. Releasing the pedal allows oil to return to the brake reservoir and more oil to enter the brake valve area past the reservoir check valve for the next braking.

On large machines, an accumulator is used as a "booster" in case the power braking fails. The accumulator holds enough "charges" of pressure oil in reserve for several brakings. When this

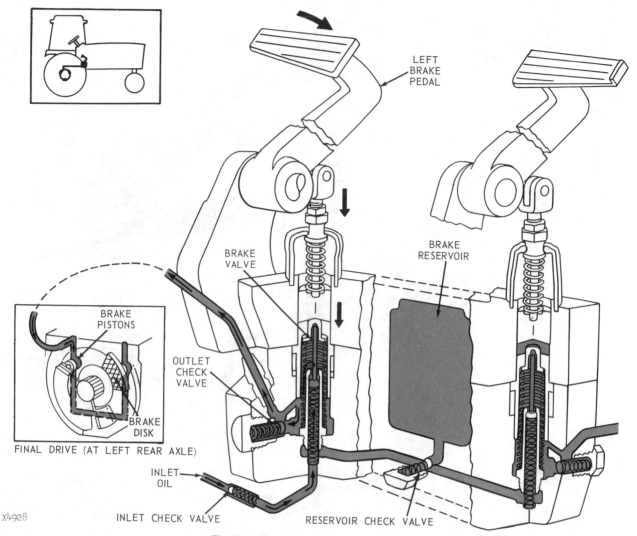

Fig. 23 — Power Brakes During A Left Turn

reservoir power is used up, the brakes can still be applied using oil trapped in the circuit.

HYDRAULIC LOAD SENSING FOR REAR-MOUNTED EQUIPMENT

On modern tractors, equipment such as a plow is often mounted on a power lift (or rockshaft) at the rear of the tractor. The equipment is controlled by hydraulics using two means:

- **Control lever**
- **Automatic load sensing**

Let's say the plow in Fig. 24 is pulling too hard and needs to be raised. We'll explain what happens when the plow is raised, first by the control lever and then by the automatic load sensing device.

Raising Plow Using Control Lever

In Fig. 24, the plow hits hard ground and the operator wants to raise it slightly. So he moves the

control lever (1) to the front. This pivots the cam follower (2) forward and presses it against rod (3) which opens valve (4). Pressure oil is now admitted to the cylinder, forcing piston (5) to the rear. The piston pushes against the shaft arm (6), rotating rockshaft (7) and lift arm (8) upward. The lift arm is attached to the plow and so the plow is raised slightly to help it pass through the hard ground.

The plow stops rising when the valve (4) is closed again, trapping oil in the cylinder. This happens when the cam follower (2) working on the sloping cam of the rockshaft (7), is returned to the rear and releases the rod (3). The valve is then closed by its spring.

Automatic Load Sensing Raises the Plow

In Fig. 25, the plow hits hard ground (1) and load control shaft (3) is pulled rearward by draft links

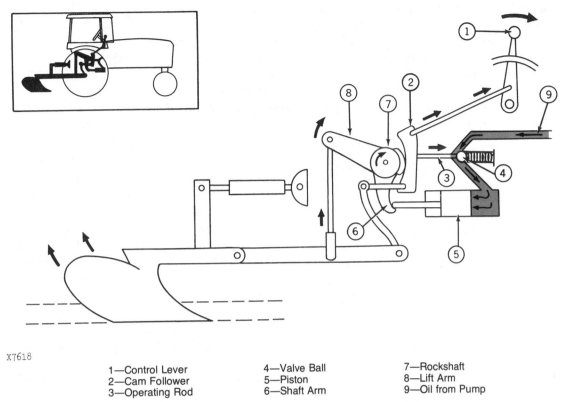

X7618

1—Control Lever	4—Valve Ball	7—Rockshaft
2—Cam Follower	5—Piston	8—Lift Arm
3—Operating Rod	6—Shaft Arm	9—Oil from Pump

Fig. 24 — Raising the Plow Using the Control Lever

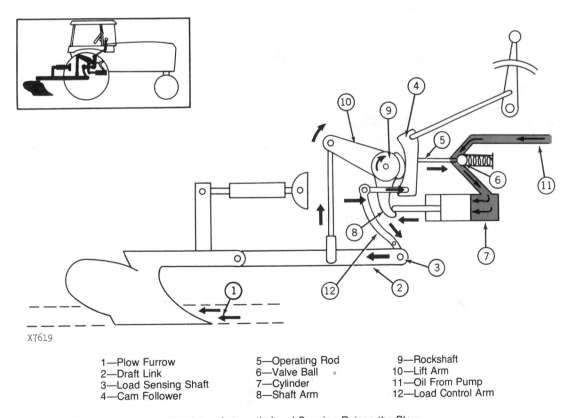

X7619

1—Plow Furrow	5—Operating Rod	9—Rockshaft
2—Draft Link	6—Valve Ball	10—Lift Arm
3—Load Sensing Shaft	7—Cylinder	11—Oil From Pump
4—Cam Follower	8—Shaft Arm	12—Load Control Arm

Fig. 25 — Automatic Load Sensing Raises the Plow

(2). As a result, load control arm (12) is pivoted against cam follower (4). The cam follower pushes on rod (5) which opens valve (6) and admits pressure oil to cylinder (7). The cylinder piston pushes against the arm (8), rotating rockshaft (9) and lift arm (10) upward. The lift arm is attached to the plow and so raises the plow slightly to help it pass through the hard ground.

With automatic load sensing, the plow will lower itself again when the hard ground is passed. This happens as the tension on the load control shaft (12) is partly released. It flexes forward and pulls back on the linkage to actuate other valves (not shown) which release some oil from the cylinder. This lets the lift arms (10) "settle," lowering the plow again.

The regular depth of the plow can be set at the control lever. The plow will stay at this depth unless a signal is given by the load sensing device.

We have just described full load sensing or *load control.*

Two other options are commonly available: If he wishes, the operator can lock out signals from the load sensing linkage by blocking the cam follower

(4) at its lower end using a lever (not shown). This is called *depth control,* since the plow now stays at the depth set by the control lever. The other choice allows the operator to partly block the cam follower using the lever. This is called *load and depth control,* since the load signals are now modified by the depth setting. On some tractors, the load and depth control is infinitely variable.

Some power lifts are designed for dual operations. Two hydraulic cylinders are normally used: one cylinder and valving controls the rear-mounted equipment, and the second operates a front-mounted tool or an attachment. However, the cylinders can be operated in parallel by opening valves which join them to the same pressure oil supply. Two control levers are also used, one for each function. But usually only one function (the rear one) has an automatic load sensing device.

HYDRAULIC SENSING ROCKSHAFT

The hydraulic load sensing system consists of a load control valve (1) and a sensing clylinder (2). The load control selector (3) is in a position at the top of the cam follower (4) to allow maximum load sensing. The rod end of the sensing cylinder piston

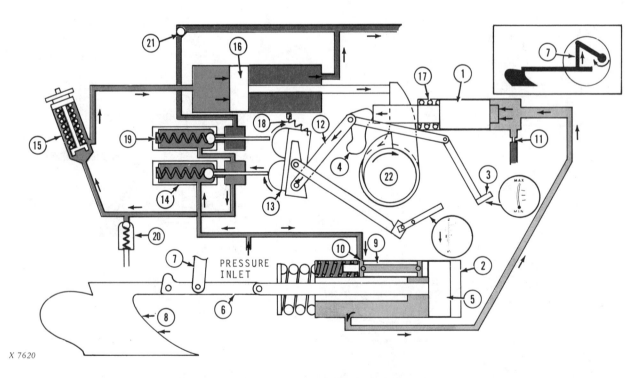

X 7620

1—Load Control Valve	6—Draft Arm	11—Orifice	17—Load Control Valve Housing
2—Sensing Cylinder	7—Draft Link	12—Valve Operating Link	18—Spring
3—Load Control Selector	8—Soil Resistance	13—Valve Operating Cam	19—Return Valve
4—Cam Follower	9—Sensing Cylinder Valve	14—Pressure Valve	20—Thermal Relief Valve
5—Sensing Cylinder Piston	10—Variable Orifice	15—Throttle Valve	21—Check Ball
		16—Rockshaft Piston	22—Rockshaft

Fig. 26 — Hydraulic Sensing Rockshaft

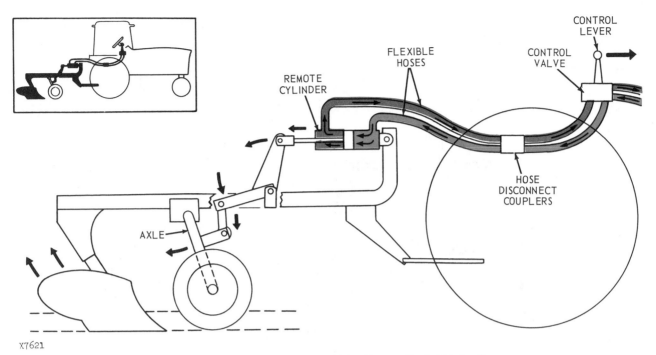

X7621

Fig. 27 — Raising A Pull-Type Plow Using Remote Control Hydraulics

(5) is attached to one draft arm (6). The two draft arms are connected by a shaft. The draft links (7) are attached to the draft arms.

As the plow hits hard ground (8), the soil resistance increases the draft load on the draft arms. The draft force is transmitted to the sensing cylinder (2) by the draft arms and pulls the sensing cylinder piston and valve (9) rearward. More oil then flows into the sensing cylinder through a variable orifice (10), causing sensing pressure on the front of the load control valve (1) to increase. The increase in sensing pressure results in rearward movement of the load control valve (1), cam follower (4), and valve operating link (12) which causes the valve operating cam (13) to rotate clockwise. Note that an orifice (11) allows some oil to escape from the front of the load control valve. This results in a variable pressure on the valve depending on the amount of oil that enters through the variable orifice (10).

The clockwise rotation of the valve operating cam causes the pressure valve (14) to open and direct pressure oil through the throttle valve (15) to the backside of the rockshaft piston (16). The throttle valve controls the speed of oil flow to and from the rockshaft piston. The piston moves forward and causes the rockshaft (22) to rotate, lifting the draft links (7) and raising the plow. The check ball (21) prevents return oil from the front of the rockshaft piston from entering the return valve housing (19).

When the plow passes the hard ground, the draft force on the sensing cylinder decreases. The sensing

cylinder piston (5) and valve (9) move forward permitting less oil to flow through the variable orifice, decreasing sensing pressure at the front of the load control valve. The spring in the load control valve housing (17) pushes the valve forward. The spring (18) causes the valve operating cam to rotate counterclockwise and forces the cam follower (4) forward, along with the load control valve. The pressure valve (14) closes and, if necessary, the return valve (19) opens to lower the plow.

When the pressure and relief valves are closed, oil trapped at the rear of the rockshaft may expand if the oil temperature rises. The thermal relief valve (20) senses thermal expansion of hydraulic oil in the system and opens if the expansion is too great.

REMOTE CONTROL OF EQUIPMENT

Tractors may operate equipment that is not mounted, but is pulled or pushed. To control this equipment with hydraulics, a remote actuator such as a cylinder or a motor is needed—separate from the tractor connected by flexible hoses.

Let's take the case of a plow again — this time one that is pulled behind the tractor (Fig. 27). The plow is pulling too hard and the operator wants to raise it. What happens is this:

The operator moves the control lever to the front as shown. This actuates the control valve which sends pressure oil to the front of the remote cylinder. As this oil pushes the piston to the rear, the cylinder rod extends. Oil from the other side of

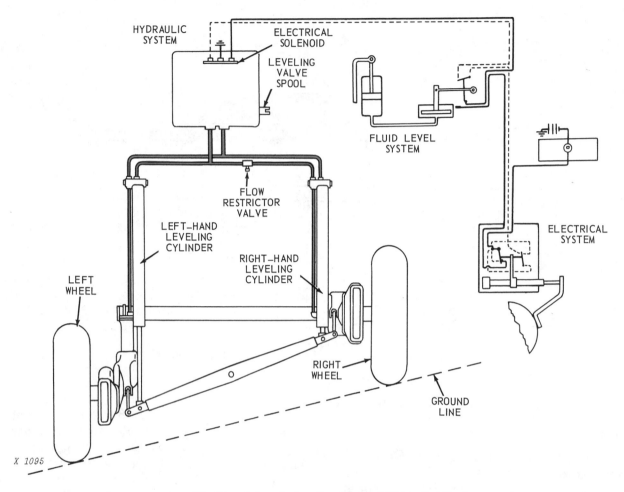

Fig. 28 — Automatic Leveling System (Hillside Combine)

the piston is forced out and returns through the valve to reservoir. As the cylinder rod extends, it pivots a linkage to the plow axle, rotating the bent axle to the rear and so raises the plow.

Remote cylinders have hundreds of uses on modern machines. (See Chapter 4 for details on all types of hydraulic cylinders.)

Another use of remote hydraulics is the hydraulic motor (see Chapter 5). A small motor can be mounted on a portable grain elevator, for example.

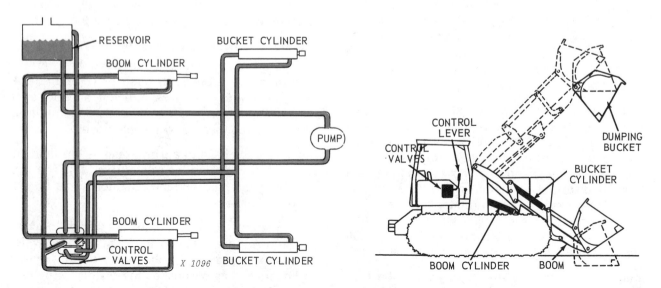

Fig. 29 — Loader Operation

The motor converts fluid power into rotary motion and so drives the elevating mechanism.

AUTOMATIC LEVELING SYSTEMS
(Hillside Combines)

A major use of the automatic leveling system is in the hillside combine.

This special system has three parts: a fluid level system, an electrical system, and a hydraulic system (Fig. 28). The first two systems provide control for activating the hydraulic system.

Operation is as follows: When the combine enters a hillside slope and left wheels lower, a liquid leveling device activates the electrical system. A solenoid moves a leveling valve spool and directs oil to double-acting leveling cylinders on each wheel (Fig. 28). The left cylinder extends and the right cylinder retracts, so that the wheels conform to the slope while the combine is held level.

When leveling in the opposite direction, the sequence is reversed.

After leveling, oil is trapped in the cylinders by check valves and the leveling system is shut off automatically.

MOUNTED EQUIPMENT CONTROL SYSTEMS

Loaders, bulldozers, backhoes, and forklifts are usually mounted on the machine which propels them. Often they are sold as custom units for a single job. For ease of control, hydraulics is widely used to operate this mounted equipment.

Loader Hydraulic Systems

Most loaders are mounted on the front of wheel or crawler tractors.

Fig. 29 shows a crawler loader which has its own hydraulic system, an open-center type.

Most loaders have two kinds of control: 1) raise and lower boom, 2) dump and retract bucket. Both are usually controlled by hydraulics, using separate oil circuits and controls.

When the operator dumps the loader bucket, he moves a control lever which causes a control valve to send oil to both bucket cylinders. This extends the cylinders, dumping the bucket. Meanwhile, oil is trapped in the boom cylinders, holding up the boom.

The boom and bucket cylinders are double-acting so that they can both raise and lower the heavy boom, or dump and retract the bucket. (Some boom cylinders may be single-acting and are lowered by the weight of the bucket.)

Each circuit—boom and bucket—is served by its own control valve. Each valve is usually operated by its own lever. In some cases, one lever is linked to both valves for four-way operation.

An extra hydraulic circuit is sometimes added for a clam bucket or a logging fork or grapple which mounts on the loader boom.

Bulldozer Hydraulic Systems

The bulldozer usually mounts on the front of a crawler tractor, for greater traction in loose dirt.

Most bulldozers have three types of blade control: 1) raise and lower, 2) angle right and left, and 3) tilt side-to-side (see Fig. 30). On some bulldozers, all three are controlled by hydraulics. On others, only one or two.

Like loaders, most bulldozers have their own hydraulic systems. Gear-type pumps, spool valves in a "stack," and double-acting cylinders are the common features of these systems.

A float position is built into the control valve which operates the raising and lowering of the blade. This allows the blade to follow the contours of the ground while backdragging and smoothing the earth. In float position, oil in the circuit is free to flow back and forth. (Normally the oil is trapped, holding the blade solid.)

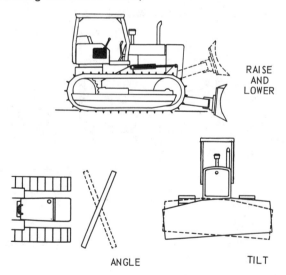

Fig. 30 — Bulldozer Operation

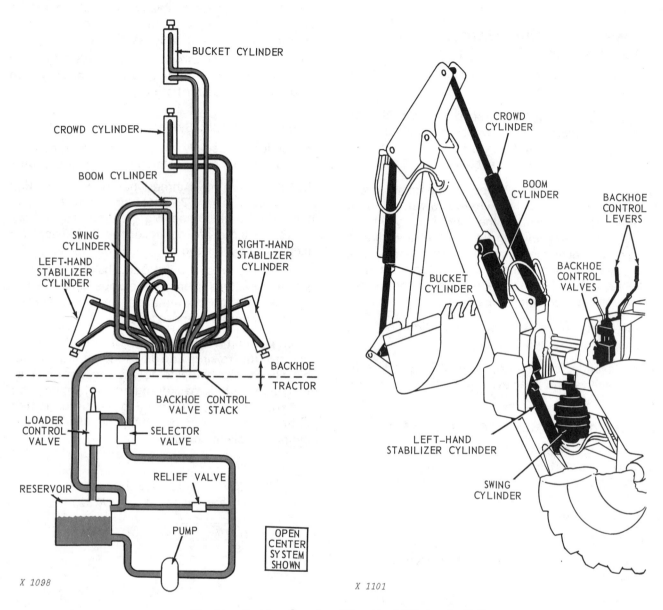

Fig. 31 — Backhoe Operation (When Used With Loader Unit)

Backhoe Hydraulic Systems

The backhoe is used for digging trenches. It usually mounts on the rear of an industrial tractor such as a loader or bulldozer.

Fig. 31 shows a typical backhoe. Hydraulic oil for the backhoe is supplied from the tractor hydraulic system. With open-center systems, a selector valve is often used to divert oil during backhoe operation. (In Fig. 31, the selector valve diverts oil from the loader when using the backhoe. With closed-center systems, oil is available "on demand.")

The operator controls the backhoe by means of levers. These levers direct oil through control valves to the proper cylinder to operate the boom, bucket, crowd, or stabilizing functions. The cylinders are double-acting to give full force in both directions. A special swing cylinder is used to rotate the boom for dumping the bucket and returning to the trench.

The use of flexible hoses shown in Fig. 31 allows free movement of the backhoe without damage to the hydraulic circuits.

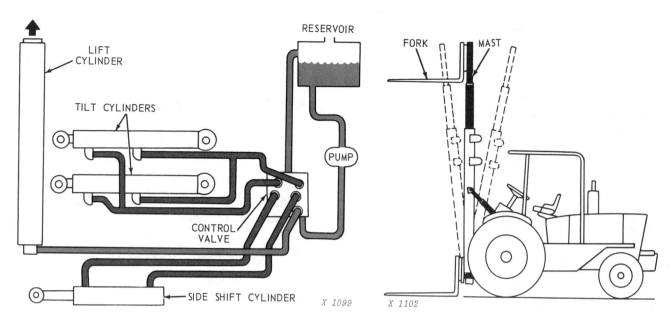

Fig. 32 — Forklift Operation

Forklift Hydraulic System

The forklift is used to handle, lift, and stack products and materials. Many forklifts are mounted on the rear of a wheel tractor. The tractor is then operated in reverse, the operator facing the forklift.

Fig. 32 shows a typical forklift. The vertical frame is called a mast, while the lifting device is called the fork. The forklift may have its own hydraulic system, either a closed- or an open-center type. Spool-type control valves and single- or double-acting cylinders are normally used in the system.

Most forklifts have three types of hydraulic control: 1) to lift and lower the fork, 2) to tilt the mast fore and aft, 3) to shift the mast from side-to-side (optional).

To raise the fork and lift a load, the operator moves a control lever to direct oil as shown in Fig. 32. The control valve sends pressure oil to the lift cylinder, while oil is trapped in the tilt and side shift circuits. (One spool valve is provided for each of the three circuits.)

DIAGNOSIS AND TESTING OF HYDRAULIC SYSTEMS

In the final chapter of this manual, we will once again return to the complete hydraulic system. At that time we will use our knowledge of "how the system works" to find out "why the system fails" and "how to remedy" these failures. Chapter 12 is titled "Diagnosis and Testing of Hydraulic Systems."

But before we try to analyze the complete hydraulic system, we must look at the various working parts in more detail. The following chapters will do that, starting with the hydraulic pump.

HYDRAULIC FACTS

Here are some key facts that will help you understand hydraulics:

1. Hydraulic power is nearly always generated from mechanical power. Example: A hydraulic pump driven by an engine crankshaft.

2. Hydraulic power output is nearly always achieved by converting back to mechanical energy. Example: A cylinder which raises a heavy plow.

3. There are three types of hydraulic energy: a) potential or pressure energy; b) kinetic energy, the energy of moving liquids; and c) heat energy, the energy of resistance to flow, or friction.

4. Hydraulic energy is neither created nor destroyed, only converted to another form.

5. All energy put into a hydraulic system must come out either as work (gain) or as heat (loss).

6. When a moving liquid is restricted, heat is created and there is a loss of potential energy (pressure) for doing work. Example: A tube or hose that is too small or is restricted. Orifices and relief valves are also restrictions but they are purposely designed into systems.

7. *Flow through an orifice or restriction causes a pressure* **drop.**

8. *Oil must be confined to create pressure for work. A tightly sealed system is a must in hydraulics.*

9. *Oil takes the course of least resistance.*

10. *Oil is normally* **pushed** *into a pump, not drawn into it. (Atmospheric pressure supplies this push. For this reason, an air vent is needed in the top of the reservoir.)*

11. *A pump does not pump pressure; it creates flow. Pressure is caused by* **resistance** *to flow.*

12. *Two hydraulic systems may produce the same power output—one at high pressure and low flow, the other at low pressure and high flow.*

13. *A basic hydraulic system must include four components: a reservoir to store the oil; a pump to push the oil through the system; valves to control oil pressure and flow; and a cylinder (or motor) to convert the fluid movement into work.*

14. *Compare the two major hydraulic systems:*

   Open-Center System = pressure is varied but flow is constant.

   Closed-Center System = flow is varied but pressure is constant.

15. *There are two basic types of hydraulics:*

   a) *Hydrodynamics is the use of fluids at high speeds "on impact" to supply power. Example: a torque converter.*

   b) *Hydrostatics is the use of fluids at relatively low speeds but at high pressures to supply power. Example: most hydraulic systems, and all those covered in this manual.*

TEST YOURSELF

QUESTIONS

1. What are the four basic principles of hydraulics?

2. How do you determine the force exerted by a piston?

3. What four components are needed to complete a very basic hydraulic system?

4. (Fill in the blanks.) "In _____-center systems, pressure is varied but flow is constant." "In _____-center systems, flow is varied but pressure is constant."

5. Describe the difference in the control valve during neutral in an open-center system, as compared to a closed-center system.

6. Is a fluid pushed into a pump or drawn into it?

7. Does a pump create pressure?

8. Oil flow through an orifice causes pressure beyond the orifice to:

   a. rise
   b. drop

HYDRAULIC PUMPS / CHAPTER 2

INTRODUCTION

The pump is the heart of the hydraulic system. It creates the flow of fluid which supplies the whole circuit.

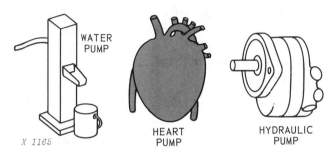

X 1165

Fig. 1—Three Kinds of Pumps

The human heart is a pump (Fig. 1). So was the old water pump once found on the farm. Somewhere in between, engineers have devised many kinds of hydraulic pumps, which do more than the old water pump, but only strive for the perfection of the human heart pump.

Once the term "hydraulics" meant the study of fluids in motion. Therefore, any pump which moved fluids was considered a hydraulic pump.

But today, "hydraulics" means the study of fluid pressure and flow—fluids in motion **plus** the ability to do work.

Thus a hydraulic pump is now one that moves fluid **and** induces fluid to work . . . in other words, A PUMP THAT CONVERTS MECHANICAL FORCE INTO HYDRAULIC FLUID POWER.

WHEN IS A PUMP "HYDRAULIC"?

All pumps create flow. They operate on a principle called **displacement**. The fluid is taken in and displaced to another point.

Displacement can be done in two ways:

• **Non-Positive Displacement**

• **Positive Displacement**

Fig. 2 compares the two. The old water wheel shows the non-positive aspect. It simply picks up fluid and moves it.

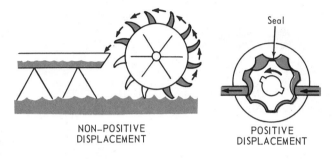

X 1166

Fig. 2—How to Tell When a Pump is "Hydraulic"

But the positive displacement pump, used in hydraulics today, not only creates flow, it also **backs it up**. Notice the sealed case around the gear. This traps the fluid and holds it while it moves.

As the fluid flows out of the other side, it is sealed against back-up. This sealing is the "positive" part of displacement. Without it, the fluid could never overcome the resistance of the other parts in the system.

When high pressure is needed in a circuit, a positive displacement pump is a must. This is true for all modern hydraulic systems which provide fluid power.

In low-pressure systems, such as water cooling or crop spraying types, the old non-positive displacement pump still works.

In this chapter, we will discuss only the **positive displacement** pump which is the heart of modern oil hydraulic systems. This pump is a true HYDRAULIC pump.

DISPLACEMENT OF HYDRAULIC PUMPS

"Displacement" is the volume of oil moved or displaced during each cycle of a pump.

In this sense, hydraulic pumps fall into two broad types:

• **Fixed Displacement Pumps**

• **Variable Displacement Pumps**

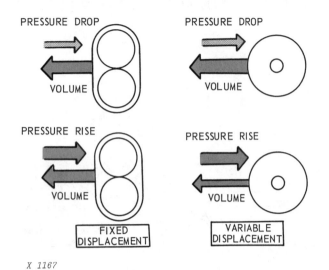

X 1167

Fig. 3—Comparing Fixed and Variable Displacement Pumps

• FIXED DISPLACEMENT pumps move the same volume of oil with every cycle. This volume is only changed when the speed of the pump is changed.

Volume can be affected by the pressure in the system, but this is due to an increase in leakage back to the pump inlet. Usually this occurs when pressure rises. This leakage means that fixed displacement pumps are usually found in lower pressure systems or as aids to another pump in a higher pressure system.

• VARIABLE DISPLACEMENT pumps can vary the volume of oil they move with each cycle—even at the same speed. These pumps have an internal mechanism which varies the output of oil, usually to maintain a constant pressure in the system. As shown in Fig. 3, when system pressure drops, volume increases. As pressure rises, volume decreases.

To summarize:

Fixed Displacement = Constant Flow

Variable Displacement = Varied Flow

In Chapter 1, we compared open center and closed center systems. We found that in open center systems, pressure is varied, but flow is constant. In a closed center system, flow is varied but pressure is constant.

Now we can see which type of pump works best for each system:

Open Center System = Fixed Displacement

Closed Center System = Variable Displacement

There are variations to this rule, as we saw in Chapter 1. But basically the two types of pumps are designed to serve the two types of systems.

A word about pressure. Remember: **A hydraulic pump does not create pressure; it creates flow. Pressure is caused by resistance to flow.**

TYPES OF HYDRAULIC PUMPS

Now that we have seen what hydraulic pumps are and what they can do, let's take an "inside" look.

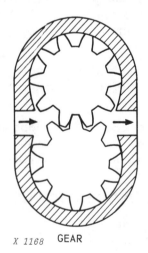

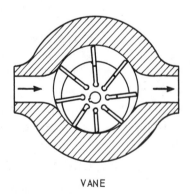

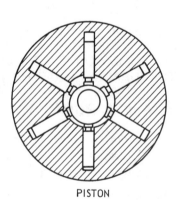

X 1168   GEAR

VANE

PISTON

Fig. 4—Three Types of Hydraulic Pumps

Most pumps used on today's machines are of three basic designs (Fig. 4):

- **Gear Pumps**
- **Vane Pumps**
- **Piston Pumps**

We will show how each type of pump operates and how it is used. A hydraulic system may use one of these pumps, or it may use two or more in combination.

All three designs work on the rotary principle: a rotating unit inside the pump moves the fluid. A rotary pump can be built very compact, yet displace the necessary volume of fluid. This is the number one need in a mobile system where space is limited.

GEAR PUMPS

Gear pumps are the "pack horses" of hydraulic systems. They are widely used because they are simple and economical. While not capable of a variable displacement, they can produce the volume needed by most systems using fixed displacement. Often they are used as charging pumps for larger system pumps of other types.

Two basic types of gear pumps are used:

- **External Gear Pumps**
- **Internal Gear Pumps**

Let's see how they work.

EXTERNAL GEAR PUMPS

External gear pumps usually have two gears in mesh, closely fitted inside a housing (Fig. 5). The drive shaft drives one gear, which in turn drives the other gear. Shaft bushings and machined surfaces or wear plates are used to seal in the working gears.

Operation is quite simple (Fig. 6). As the gears rotate and come **out** of mesh, they trap inlet oil between the gear teeth and the housing. The trapped oil is carried around to the outlet chamber. As the gears mesh again they form a seal which prevents oil from backing up to the inlet. The oil is forced out at the outlet port and sent through the system.

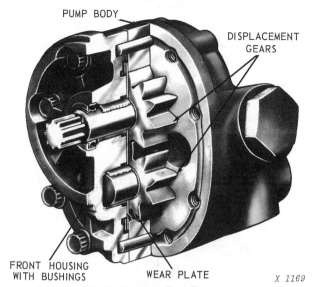

Fig. 5—External Gear Pump

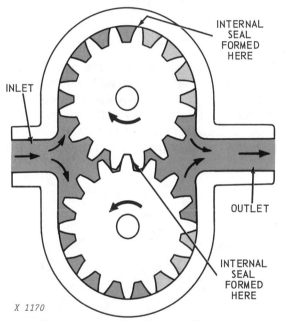

Fig. 6—External Gear Pump in Operation

This oil is pushed out by the continuous flow of trapped oil coming into the outlet chamber with each rotation of the gears.

At the inlet side, gravity feeds in more oil from the reservoir to replace that drawn out by the turning gears.

Some gear pumps use a pressurized plate working against the gears to increase pump efficiency. A small amount of pressure oil is fed under the backing plate, pressing it against the gears and forming a tighter seal against leakage.

INTERNAL GEAR PUMPS

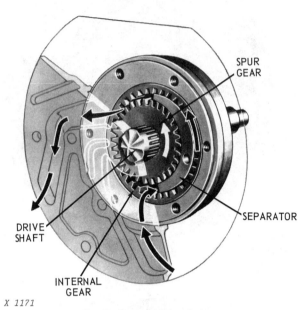

X 1171

Fig. 7—Internal Gear Pump

The internal gear pump also uses two gears, but now a spur gear is mounted **inside** a larger gear. The spur gear is in mesh with one side of the larger gear, and both gears are divided on the other side by a crescent-shaped separator. The drive shaft turns the spur gear, which drives the larger gear.

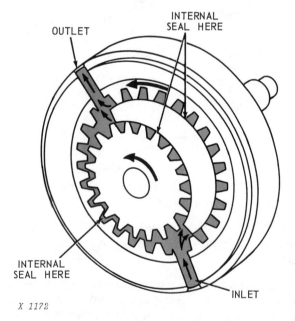

X 1172

Fig. 8—Internal Gear Pump in Operation

Operation is basically the same as for the external gear pump. The major difference is that both gears turn in the same direction (Fig. 8).

As the gears come out of mesh, oil is trapped between their teeth and the separator and is carried around to the outlet chamber. As the gears mesh again, a seal is formed, preventing back-up of the oil. A continuous flow of oil to the outlet pushes the fluid out into the circuit.

Gravity keeps feeding oil into the pump inlet to fill the partial vacuum created as oil is drawn in by the gears.

Rotor Version of Internal Gear Pump

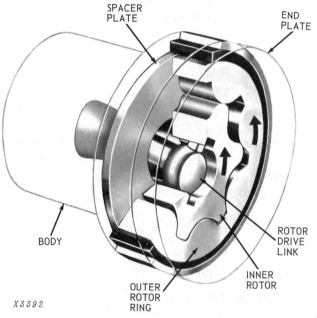

X3392

Fig. 9—Rotor Version of Internal Gear Pump

The rotor pump (Fig. 9) is a variation of the internal gear pump. An inner and outer rotor turns inside a housing. The rotor has rounded lobes for teeth. No separator is used.

In operation (Fig. 10), the inner rotor is driven inside the rotor ring. The inner rotor has one less lobe than the ring, so that only one lobe is in full engagement with the outer ring at any one time. This allows the other lobes to slide over the outer lobes, making a seal to prevent back-up of oil.

As the lobes slide up and over the lobes on the outer ring, oil is drawn in. As the lobes fall into the ring's cavities, oil is squeezed out.

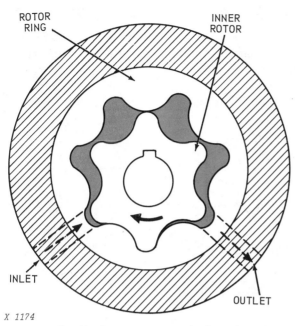

Fig. 10—Rotor-Type Pump in Operation

VANE PUMPS

Vane pumps are fairly versatile pumps and can be designed as single, double, or even triple units.

All vane pumps move oil using a rotating slotted rotor with vanes fitted into the slots.

Two types of vane pumps are most often used:

• **Balanced Vane Pumps**

• **Unbalanced Vane Pumps**

The balanced vane pump is strictly a fixed displacement type. The unbalanced vane can have a fixed or a variable displacement.

BALANCED VANE PUMPS

In the balanced vane pump (Fig. 11), the rotor is driven by the drive shaft and turns inside an oval rotor ring.

The vanes are fitted into the rotor slots and are free to move in or out.

The "balanced" part of this pump is shown by the position of the oil ports in Fig. 12. The pump has two inlet ports, located opposite each other. And it has two outlet ports, also on opposite sides of the pump. Both sets are connected to a central inlet and outlet.

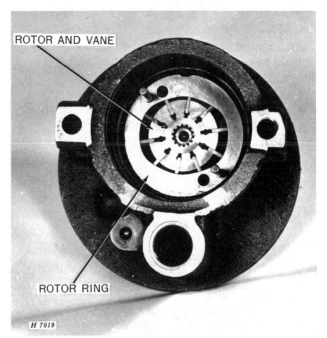

Fig. 11—Balanced Vane Pump

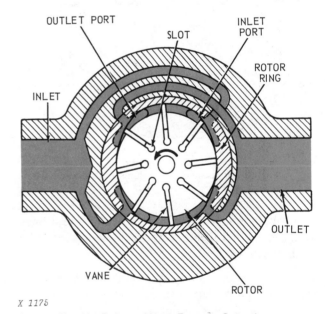

Fig. 12—Balanced Vane Pump in Operation

Operation is shown in Fig. 12. As the rotor turns, the vanes are thrown out against the inside surface of the ring by centrifugal force. As the vanes follow the contour of the oval-shaped ring, they divide the crescent-shaped areas between the rotor and the ring into two separate chambers. These chambers are continually expanding and shrinking in size—twice during each revolution. The inlet ports are located where each chamber begins to expand; the outlet ports are located where each chamber begins to shrink.

As the chamber begins to expand, inlet oil rushes in to fill the partial vacuum. This oil is carried around by the vanes. As the oil chamber begins to reduce, the confined oil is forced out at the outlet port.

In the second half of the revolution, this action is repeated at the second set of inlet and outlet ports.

UNBALANCED VANE PUMPS

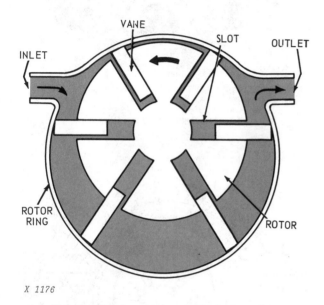

X 1176

Fig. 13—Unbalanced Vane Pump in Operation

The unbalanced vane pump uses the same basic principle of a turning rotor with vanes working inside a fixed rotor ring.

However, the operating cycle only happens once each revolution (Fig. 13). So this pump has only one inlet and one outlet port. Also, the slotted rotor is now set offside in a circular ring.

In operation, the oil chamber starts to expand at the inlet port, and finishes its contracting at the outlet port.

Oil is drawn in by the partial vacuum, and forced out by the shrinking of the chamber, the same as in the balanced vane pump.

However, the design of the unbalanced vane pump is different from the balanced type, as we'll explain now.

BALANCED VS. UNBALANCED VANE PUMPS

The balanced vane pump is really a refinement of the unbalanced model. Why was this refinement needed?

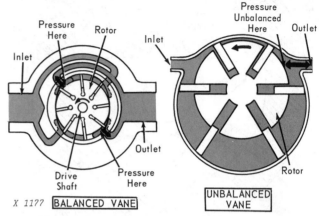

X 1177 BALANCED VANE   UNBALANCED VANE

Fig. 14—Balanced Vs. Unbalanced Vane Pumps—
Pressure on Rotor and Shaft

The answer is shown in Fig. 14. The unbalanced vane pump seemed to have frequent bearing failures. The cause was found to be force on the shaft and bearings of the back pressure from oil being expelled at the outlet side of the pump. No equal force was exerted on the opposite side, since the inlet oil was under little or no pressure.

The balanced vane pump was a solution to this problem. To balance off the outlet pressures on the shaft, **two** outlet ports were used, directly opposite each other. This equalized the forces, increased bearing life, and made the pump work longer.

While the balanced vane pump solved one problem, it posed another one: it could only be used for fixed displacement. The outlet port positions cannot be changed or the balance would be upset.

The unbalanced model can be used either for fixed or variable displacement. By special design, the position of its rotor ring and oil ports can be changed in relation to the offset of the rotor. This changes the size of the chambers which the vanes create, thus the amount of oil each carries. The result: a variable displacement pump.

So the two vane pumps give you a choice:

1. Longer service life, or

2. More flexible operation.

The final choice for any hydraulic system depends upon the job to be done.

PISTON PUMPS

Piston pumps are often favored on modern hydraulic systems which use high speeds and high pressures.

However, piston pumps are more complex and more expensive than the other two types.

Piston pumps can be designed for either fixed or variable displacement.

Most piston pumps are included in two types:

• **Axial Piston Pumps**

• **Radial Piston Pumps**

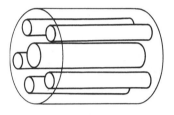

AXIAL PISTON   RADIAL PISTON

X 1178

Fig. 15—Axial and Radial Piston Principles

AXIAL piston means that the pistons are mounted in lines parallel with the pump's "axis" (a line down the center). See Fig. 15.

RADIAL piston means that the pistons are set perpendicular to the pump's center like the sun's rays.

Both styles of piston pumps operate using pistons which pump oil by moving back and forth in cylinder bores. (Another term for this movement is "reciprocate.")

The basic reciprocating piston pump is shown in Fig. 16. The most efficient pump in hydraulics, this pump is not used in machine hydraulic systems because it takes up too much space.

Axial and radial piston pumps use reciprocating pistons but drive them by the rotary principle. In this way the efficiency of the reciprocating method is combined with the compactness of the rotary operating pump.

The result is a pump which is efficient, yet can fit into a machine hydraulic system.

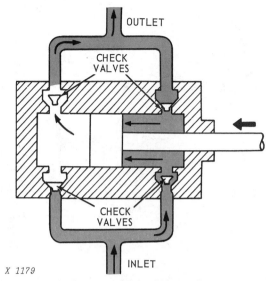

X 1179

Fig. 16—Reciprocating Piston on Pump

AXIAL PISTON PUMPS

Axial piston pumps usually fall into two broad types: inline and bent-axis.

Inline Axial Piston Pumps

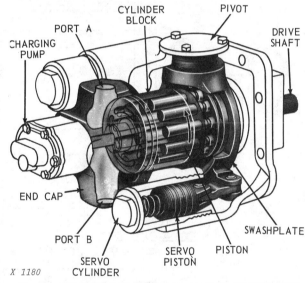

X 1180

Fig. 17—Inline Axial Piston Pump—Variable Displacement

In this pump, the cylinder block is mounted on a drive shaft and rotates with the shaft (Fig. 17).

The pistons work in bores in the cylinder block which are parallel to the axis of the block. The heads of the pistons are in contact with a tilted plate called a swashplate.

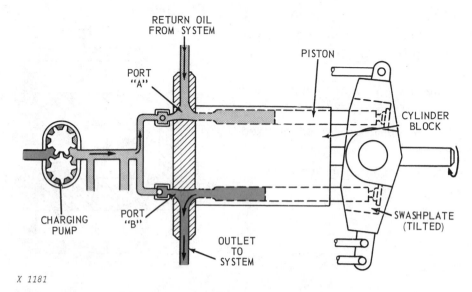

X 1181

Fig. 18—Inline Axial Piston Pump in Operation

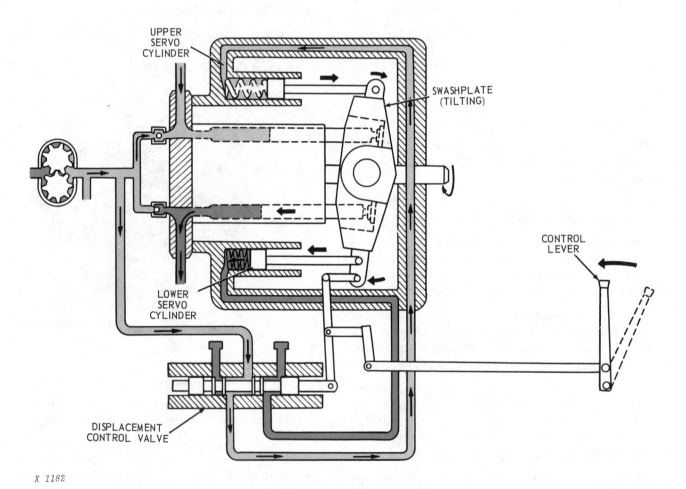

X 1182

Fig. 19—Servo Device in Operation, Tilting Swashplate

The swashplate does not turn but it can be tilted back and forth. It mounts on a pivot and is controlled either manually or by an automatic "servo" device.

Since the swashplate controls the output of the pistons, this pump has a variable displacement.

First let's see how the basic pump works and then we'll explain the servo device.

In Fig. 18 the swashplate has been tilted to the right at the top.

Remember that the angle of the swashplate controls the distance that the pistons can move back and forth in their bores. The greater the angle, the farther the pistons travel and the more oil that is displaced by the pump.

When the swashplate is tilted as shown, port "A" is the inlet port. As the cylinder block rotates, piston bores align with this port and oil is forced into the bores by the small charging pump. This oil pushes the pistons against the swashplate. Then as they revolve, these pistons follow the tilt of the swashplate and force the oil out of their bores into port "B", the outlet.

If the angle of the swashplate was fixed, the pump would operate as a fixed displacement type, putting out the same amount of oil with each revolution.

But on this pump the swashplate can be moved—hence, variable displacement. This is done by a servo device as we'll explain now.

The servo device has been added in Fig. 19.

To tilt the swashplate, the control lever is actuated, moving the displacement control valve to the left. This directs oil from the charging pump into the upper servo cylinder, moving the piston which tilts the swashplate.

Meanwhile the piston in the lower servo cylinder is pushed in by the lower part of the swashplate, forcing its oil back through the valve to the pump case.

When the swashplate reaches the angle set by the control lever, the control valve returns to neutral and traps the oil in the servo cylinders. This holds the swashplate until the control lever is moved again.

The pump keeps on pumping as explained before, drawing in oil at the top and pushing out oil at the bottom of each revolution.

If the swashplate were tilted the opposite way, the inlet-outlet cycle of the pump would be reversed. Oil would be drawn in at the bottom and pushed out at the top. So the servo device not only controls the pump displacement but also the direction of this oil.

Because the pump provides for variable displacement and direction control, it is widely used for hydrostatic drive trains.

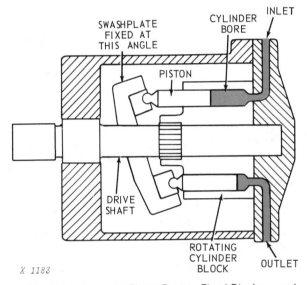

X 1183

Fig. 20—Inline Axial Piston Pump—Fixed Displacement

There are three other types of inline piston pumps. Two are fixed displacement types, one is a variable displacement pump. The first pump, Fig. 20, operates on the same principle we have just described. The only difference is that no servo device is used. The swashplate is fixed at the angle shown, therefore it has fixed displacement. The second pump, Fig. 21, also operates on the swashplate principle but it provides variable displacement. The swashplate angle can be changed either mechanically or hydraulically. The change in swashplate position governs the amount of oil being pumped. Because oil flows only in one direction, this pump is referred to as a single direction, variable displacement axial piston pump.

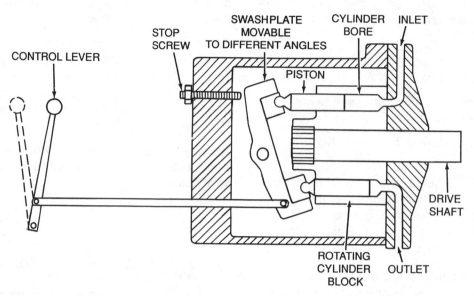

Fig. 21 — Inline Axial Piston Pump — Variable Displacement

The third pump, as you can see in Fig. 22 is quite different. The cylinder block is stationary while the swashplate rotates. The pistons contact the swashplate as it rotates and slide the pistons back and forth in their bores, pumping oil.

Valves are used to separate inlet and outlet oil routes to each piston bore as shown. The check valve balls seat to prevent oil from going out the outlet gallery until forced out by the piston.

Each piston acts like a separate pump, opening and closing its valves to cycle its oil with each revolution.

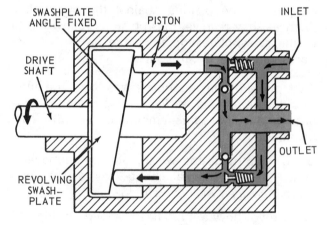

X 1184

Fig. 22 — Inline Axial Piston Pump — Fixed Displacement

Bent-Axis Axial Piston Pump

Another type of axial piston pump is the bent-axis type. Fig. 23 shows a fixed displacement model of this pump.

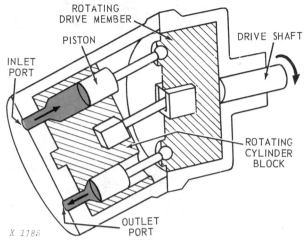

Fig. 23 — Bent-Axis Axial Piston Pump — Fixed Displacement

In this pump, the pump housing is slanted in relation to the driving member. The piston heads are connected to the drive member, which is driven by the drive shaft.

Both the cylinder block and drive member rotate and are enclosed in the pump housing.

As the two units revolve, the pistons are forced in and out of their bores by the tilting of the drive member, pumping oil as shown.

RADIAL PISTON PUMPS

Radial piston pumps are among the most sophisticated of all pumps. They are capable of high pressures, high volumes, high speeds, and variable displacement.

The basic operation is simple, but by using extra valves and other devices, this pump can be adapted to many systems and needs.

This pump is closely fitted, so wear can be a problem unless clean oil is used. And the oil must contain properties which lubricate the closely fitted parts.

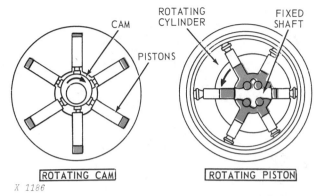

Fig. 24 — Two Operating Principles of Radial Piston Pumps

Radial piston pumps are designed to operate in two ways (Fig. 24).

In the "rotating cam" pump, the pistons are located in a fixed pump body. The center shaft has a cam which drives the pistons as it rotates.

In the "rotating piston" pump, the pistons are located in a rotating cylinder. As the cylinder rotates, the pistons are thrown out against the outer housing. Since the rotating cylinder is set offside in the housing, the pistons are moved back and forth as they follow the housing.

Let's follow the basic operation of each pump.

Radial Piston Pump (Rotating Cam Type)

The typical radial piston pump shown in Fig. 25 uses the "rotating cam" principle and is normally designed as a four- or eight-piston model.

Radial piston pumps can also be designed as dual bank pumps which doubles the displacement, but still keeps the pump size relatively small in relationship to capacity. The dual bank pump only requires one common drive shaft.

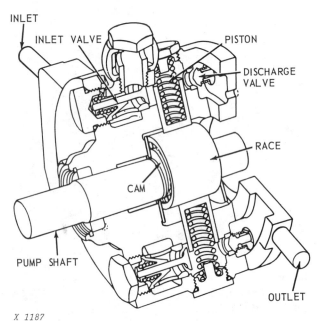

X 1187

Fig. 25 — Radial Piston Pump (Rotating Cam Type)

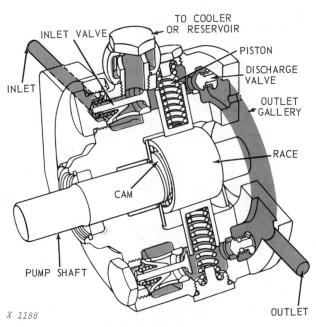

X 1188

Fig. 26 — Radial Piston Pump in Operation
(Without Stroke Control Mechanism)

The radial pistons are located in bores in a fixed housing. The drive shaft has an eccentric cam which contacts the pistons as it turns, moving them out to pump the oil.

Oil inlet and outlet is through annular passages, at each end of the pump housing. Ports on each side of the piston bore connect with these passages. Spring-loaded valves in the ports allow oil to flow in and out of the piston bores.

The pistons are driven outward to discharge by the shaft cam; they move inward to take in oil by force of their springs.

As it stands now, this pump could have a fixed displacement. But if that was all we needed, a cheaper gear or vane pump would do the job as well or better.

The radial piston pump is used only where extra features like variable displacement are required.

For variable displacement, a stroke control mechanism is used.

To understand how the stroke control feature works, let's first look at how the pump works without it—as a fixed displacement type.

INLET STROKE OF PISTONS

When the piston springs return the piston to the center of the pump, a partial vacuum is created in the bore. This vacuum plus oil pressure opens the inlet valve and incoming oil fills the piston bore. When the bore is filled, the vacuum is gone and the inlet valve is closed by its spring.

A small charging pump is commonly used to feed in low pressure oil to the radial piston pump.

DISCHARGE STROKE OF PISTONS

To discharge oil, the revolving cam contacts the piston, forcing it outward. This force opens the outlet valve and discharges the oil into the outlet gallery. When the piston reaches the top of its stroke, the flow stops and the outlet valve is closed by its spring.

The piston then starts on its inlet stroke and the cycle begins again.

The cycle of each piston works in rapid sequence as the cam rotates. This produces a constant flow of oil.

The output of oil depends on the speed of the pump alone—**if** it were a fixed displacement pump.

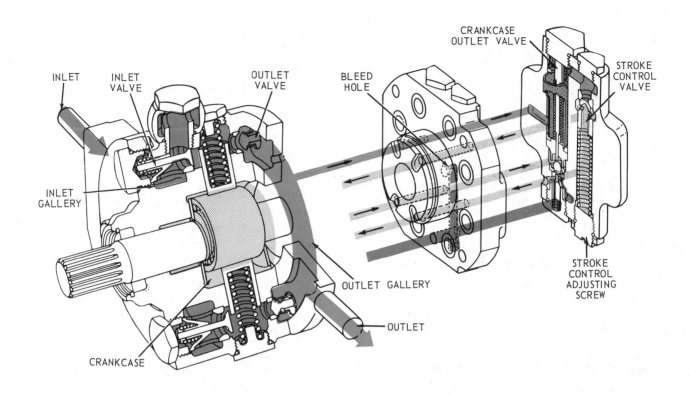

INLET

INLET
VALVE

OUTLET
VALVE

BLEED
HOLE

CRANKCASE
OUTLET VALVE

STROKE
CONTROL
VALVE

INLET
GALLERY

OUTLET GALLERY

STROKE
CONTROL
ADJUSTING
SCREW

OUTLET

CRANKCASE

X 1189

Fig. 27 — Radial Piston Pump Using Stroke Control Mechanism

USE OF STROKE CONTROL MECHANISM

How does a stroke control mechanism vary the displacement of this pump?

One way would be to slow down or stop the pump drive shaft. But this would require a manual or mechanical device that means reaction time lags and human errors.

Another way to control output would be to hold the pistons away from the driving cam. This is what the stroke control mechanism does—automatically, using hydraulics.

The stroke control valve admits oil into the crankcase at the center of the pump (Fig. 27). This oil is under sufficient pressure so that it holds the pistons away from the cam. The crankcase outlet valve is closed, trapping this pressure oil.

This causes the pump to reduce its output to nearly zero, even though the drive shaft keeps on turning.

When there is a demand from a hydraulic function, the pressure drops at the crankcase outlet valve,

allowing the spring to open it. This releases the oil from the pump crankcase into the inlet gallery.

As pressure drops in the pump crankcase, the pistons contact the cam again and start pumping oil. The pump goes back "in stroke."

As the demand for oil in the system is satisfied, the flow of oil slows down and back pressure closes the crankcase outlet valve and opens the stroke control valve. The pump keeps on pumping until the crankcase pressure holds the pistons away from the cam. The pump goes "out of stroke."

While in standby the pump still moves a slight amount of oil. This oil is directed through a bleed hole back to the inlet gallery to cool and lubricate the pump.

Initial pressure in the system is controlled by an adjusting screw on the stroke control valve. This screw controls the system pressure at which oil is released into the pump crankcase to build pressure. By controlling the distance that the pistons can travel toward the cam, this controls the amount of fluid that flows into the piston bores to be pumped to the system.

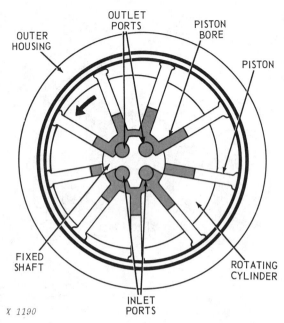

OUTLET
PORTS

PISTON
BORE

OUTER
HOUSING

PISTON

FIXED
SHAFT

ROTATING
CYLINDER

INLET
PORTS

X 1190

Fig. 28 — Radial Piston Pump (Rotating Piston Type)

Radial Piston Pump (Rotating Piston Type)

The other version of the radial piston pump is shown in Fig. 28.

It has rotating pistons and operates much like an unbalanced vane pump. As the eccentric cylinder rotates, the pistons work back and forth on the inner surface of the housing. The inlet and outlet ports are divided by a spindle on the fixed center shaft.

As the cylinder rotates, the pistons are thrown out against the housing by centrifugal force. A partial vacuum is created in the piston bores and oil flows in the inlet ports to fill the bores as shown.

As the cylinder keeps turning, the pistons are pushed back into the bores and force the oil out the ports on the outlet side.

Variable displacement is controlled by adjusting the relation of the outer housing to the cylinder. This governs the travel of the pistons and so the amount of oil pumped during each cycle.

This concludes our descriptions of the three types of pumps used in modern hydraulic systems.

SUMMARY OF PUMP TYPES

Before going into the application and efficiency of hydraulic pumps, let's review some of the points that we have just covered.

In summary:

1. A hydraulic pump converts mechanical force into hydraulic or fluid power—in other words, induces fluid to work.

2. Of the two main types of pumps, positive displacement and non-positive displacement, the positive displacement type is best suited for power hydraulics due to its ability to produce a steady flow against the high pressures in the system.

3. A hydraulic pump can be designed to produce either a specific volume of fluid at a specific speed, or to produce a variable volume of fluid at a constant speed . . . fixed displacement or variable displacement.

4. The three types of pumps most often used in machine hydraulic systems are gear, vane, and piston.

5. These three pumps operate on a rotary principle. This allows them to be constructed as small units, yet still have the ability to produce the required volume of fluid.

6. The preceding text covers only basic hydraulic pumps and there are a great number of variations on all of the pumps selected.

HYDRAULIC PUMP EFFICIENCY

Thus far, we have only described the three most popular types of pumps. This, of course, is not the whole story on hydraulic pumps. Their application and efficiency is just as important as their operation and this may help later in diagnosing hydraulic problems.

PUMP QUALITIES

Because of the wide variety of pumps and hydraulic systems, we could not possibly prescribe a particular model of pump for a particular system without having full information about the system.

However, we can describe the desirable and undesirable qualities of the three pumps and let you judge for yourself the reasons why a particular pump is used in a hydraulic system.

PHYSICAL SIZE

One of the first factors to consider when choosing a pump for a machine system is the pump's physical size. Most of these systems have little or no room to spare and may allow just a small area for the pump. Fortunately, with the wide variety of pumps and pump sizes available, this is not a big problem . . . unless the system requires a function that a pump cannot provide except as a large unit. In this case, space for the pump, regardless of size, will be made available, because the other more important requirements of the system cannot be sacrificed.

PUMP DELIVERY, PRESSURE, AND SPEED

One other requirement is the volume of fluid that the pump produces.

Most pumps are rated by volume, which is usually expressed in gallons per minute (gpm). This rating is called several names — delivery rate, discharge, capacity, or size. Regardless of the rating, it cannot stand alone. It must be accompanied by a figure stating the amount of back pressure that the pump can withstand and still produce the gpm rating; for as pressure increases, internal pump leakage increases and usable volume decreases.

Pump speed must also be included with the volume rating for two reasons.

First, in a fixed displacement pump, flow is directly related to the speed of the pump—the faster the speed, the more fluid pumped.

Second, how fast the pump must go to produce a certain flow indicates at what speed the driving mechanism for the pump must travel (in revolutions per minute or rpm). Add this to the delivery rate of a pump, and here is an example of how a rating could read: "11.5 gpm with 2000 psi at 2100 rpm."

Occasionally, a pump will have an alternate delivery rate, referred to as an intermittent delivery rate. This rating indicates the highest level a pump can operate, in terms of delivery, pump speed and pressure, for a short period of time and still maintain satisfactory service life.

PUMP EFFICIENCY

The efficiency of a pump (how well it does its job) is also important in selecting a pump.

We may have a pump that meets the delivery requirements of a system under the existing pressure in the system at the speed that is available to drive the pump—we may have all this and more . . . but what if we find that the pump requires a great amount of mechanical power to attain this delivery rate? Or what if we find that the materials in the pump must be specially and expensively constructed to withstand the pressure or friction in

the system? This is why the knowledge of pump efficiency is important before selecting a particular pump. We are not only looking for delivery rate, but delivery rate provided by efficient and economical operating means.

Pump quality is judged by three ratings:
- **Volumetric Efficiency**
- **Mechanical Efficiency**
- **Over-All Efficiency**

VOLUMETRIC EFFICIENCY is the ratio of the actual output of the pump to theoretical output (the amount it should put out under ideal conditions). The difference is usually due to internal leakage in the pump.

MECHANICAL EFFICIENCY is the ratio of the over-all efficiency of the pump to volumetric efficiency. This difference is usually due to wear and friction on the pump's working parts.

OVER-ALL EFFICIENCY is the ratio of the hydraulic power **output** to the mechanical power **input** of the pump. This is the product of both mechanical efficiency and volumetric efficiency.

SUMMARY OF PUMP SELECTION

There are sundry other factors in judging pumps and pump applications—such as a pump's adaptability to certain fluids—its adaptability to different types of systems and system layouts—the environment that the pump will be working in—cost of the pump—etc.

All of these factors have a part in selecting a pump for a particular system.

CLASSIFYING GEAR, VANE, AND PISTON PUMP EFFICIENCY

Now that we have discussed some of the factors used in evaluating a hydraulic pump, let's compare the three types of pumps and see how they rate.

Remember, though, that we are still talking in a general sense. For more detailed specifications, go to the machine technical manual.

PHYSICAL SIZE

In physical size, you will find that all three types of pumps will range from very small to very large. Physical size, then, is not too important.

Of the three, you will generally find that the gear pump is the smallest, the piston pumps the largest, and the vane pumps in between. The reasons for this are not so much because of the lack of space, but because of the delivery requirements of the different systems.

PUMP DELIVERY, PRESSURE, AND SPEED

Delivery rate is another matter. Piston pumps usually deliver more fluid under greater pressure and operate at higher speeds. Vane pumps are second and gear pumps are third in this respect.

Overall, this is how the three pumps measure up in the delivery rate, pressure and speed departments:

	Delivery (Gpm)	Pressure (Psi)	Speed (Rpm)
Gear Pump	0.2 - 150	250 - 2500	800 - 3500
Vane Pump	0.5 - 250	250 - 2500	1200 - 4000
Piston Pump	0.5 - 450	750 - 5000	600 - 6000

As you can see, there is a wide range of available delivery rates. This is not to say that most machines require this wide range.

Delivery rates of pumps used on modern farm and industrial equipment generally range from one to fifty gpm.

Pressure usually runs from 100 to 2500 psi. Top pump speed will generally range from 800 to 3500 rpm.

Efficiency of the gear, vane, and piston pumps range from 75% to 95%.

Piston pumps are usually rated on the high side, gear pumps on the low side, and vane pumps in the middle. These figures are based only on the three efficiency ratings. It does not include the adaptability of the pump to the system, the material used in pump construction, or the initial cost of the pump.

SUMMARY OF PUMP EFFICIENCY

In summary, the major factor in adapting a pump to a particular system is that system's overall needs. It would be foolish to use a pump with high delivery in a system that requires only a low delivery rate. By the same token, using a pump that must produce at its peak continuously just to meet the minimum requirements of the system is equally wrong. Making either of these mistakes produces a poor system due to excessive initial pump costs or continual costly repairs.

Use a pump that is suited to the system, whether a gear pump, which has fewer moving precision parts, or a piston pump, which has many parts fitted to close tolerance and is therefore more expensive.

This kind of choice will give better and more economical service than using a pump that is "overpowered" or "underpowered."

MALFUNCTIONS OF PUMPS

THE HUMAN ERROR FACTOR

The majority of hydraulic pump failures are due to human factors: poor maintenance, bad repair, exceeding operating limits, and the greatest cause—the use of fluid which is dirty or of poor quality.

Hydraulic pumps can wear out through normal use, but few pump failures can be attributed to "old age."

From the list of causes, you will see what is responsible for the majority of failures—the HUMAN ERROR factor.

To prevent this, know your hydraulics, maintain the system, operate it as designed, and use the proper fluids.

We might point out that this is a prime example of why this manual was written. By presenting you with the basic theory and operation of hydraulic systems, we hope to reduce this large percentage of human error.

In the remainder of this chapter, we will cover the general causes of pump failures, how they can be prevented, and some of the general repairs that can be made on them.

CONTAMINATED FLUID

Since hydraulic fluid and its proper use is covered in Chapter 10, we will only relate here what happens to a pump if the fluid is contaminated or will not adapt to the requirements of the pump.

As we said before, contaminated or improper fluid is the biggest offender in hydraulic pump failures.

X 1191

Fig. 29 — Pump Pistons Scored by Contaminated Fluid

Contaminated fluid can damage a pump in many ways. Solid particles of dirt, sand, etc. in the fluid act as an abrasive on the pump's closely fitted parts (Fig. 29). This causes abnormal wear on the parts and increases internal leakage, thus lowering the pump's efficiency and increasing oil temperature.

X 1192

Fig. 30 — Vane Pump Rotor Damaged by Contaminants in Fluid

Sludge is formed by the chemical reaction of fluid to excessive temperature change or condensation. It will build up on the pump's internal parts and eventually plug the pump (Fig. 30). If the pump is plugged on the inlet side, it will be starved for fluid and the heat and friction will cause the pump parts to seize.

X 1193

Fig. 31 — Abnormal Wear Created by Sludge (Vane Pump Rotor Ring)

Air, plus water, plus heat will also create sludge by oxidation and the results to the pump will be the same. See Fig. 31.

X 1194

Fig. 32 — Pump Drive Cam Pitted by Rust

Water or other foreign liquids in the fluid can rust the pump parts and housings (Figs. 32 and 33). Rust will not only build up on the metal, but will also flake off into the fluid as abrasive solid particles.

X 1195

Fig. 33 — Balanced Vane Pump Rusted by Water in the Fluid

IMPROPER FLUID

The selection of the proper fluid viscosity is extremely important. Viscosity is the rating given to the degree that the fluid resists flow. A high viscosity number indicates that the fluid is "heavy" and will strongly resist flow. A lower viscosity means a "lighter" fluid which flows faster.

Below are some of the things that may occur if the fluid viscosity is wrong:

If the fluid is too "light":

1. Both internal and external leakage will increase.

2. Oil slippage past pump parts will increase, which will cause heat and reduce efficiency.

3. Parts wear will increase for lack of adequate lubrication.

4. System pressure will be reduced.

5. Overall control of the system functions will be "spongy."

If the fluid is too "heavy":

1. Internal friction will increase, which, in turn, will increase flow resistance through the system.

2. Temperature will increase, thus increasing the chance of sludge build-up.

3. Operation of functions will be sluggish and erratic.

4. Pressure drop throughout the system will increase.

5. More power will be required for operation.

X 1196

Fig. 34 — Vane Pump Ring Wear from
Use of Improper Fluid

SOME FLUIDS—BAD FOR SEALS

The chemicals that make up some fluids can also be troublemakers. They may react unfavorably with the fibers or synthetic materials used in the packings and seals of the pump. This can lead to rapid deterioration of the seals and packings and both internal and external leakage may result. Or, the seals and packings may swell or shrink with the same result.

SOME FLUIDS—BAD IN THEMSELVES

Some fluids are not adaptable to the environment that the pump must work in. For instance, if the pump must work under constant high temperatures at peak performance, the fluid may have to do more work than it is capable of doing.

Thus a fluid can be literally worked to death. This results in many of the failures we have already mentioned.

POOR OPERATING PROCEDURES

Although only a small percentage of failures are due to exceeding the operating limits, this is still a problem that can be eliminated.

X 1197

Fig. 35 — Broken Pump Drive Shaft — Fatigue Failure
from Too-Tight Drive Belts

Remember that a hydraulic pump is a precise and delicate instrument. Overspeeding or overloading it will upset the balance between long life and high efficiency, resulting in poor operation and a damaged pump. See fatigue failure Fig. 35.

Overloading The Pump

Most pumps are rated far below their utmost capacity to provide continuous service. This allows them to be pushed up to their peak for short periods of time and still have a satisfactory operating life.

What if you continuously operated the pump at its upper limit? What would happen to the bearings within the pump, for instance?

Let's take an example: Suppose that the pump has been operating at 1000 psi and has a bearing life of 4800 hours. Now let's raise the operating pressure to 2000 psi. How long could we expect the bearings to last? There is a formula that we can use to determine this.

The formula:

$$\text{Bearing Life} = \frac{\text{Old Bearing Life}}{(\text{New Pressure/Old Pressure})^3}$$

Expressed in our example it reads:

$$\frac{4800 \text{ hours}}{(2000 \text{ psi}/1000 \text{ psi})^3} = \frac{4800}{2^3 \text{ (or } 2 \times 2 \times 2)} = \frac{4800}{8}$$

$$= 600 \text{ hours}$$

So we can see that by just doubling the pump operating pressure we have reduced the life of the bearings and subsequently the life of the pump by eight times (from 4800 hours to only 600 hours).

In operation, many things can cause the pump pressure to rise: poor maintenance, overloading the pump, hidden restrictions in the system.

Overspeeding the pump

By increasing the pump speed, we can also reduce bearing life. Here, again, we have a formula to show this. Let's take the same pump we used above and double its speed. Remember that the pump has a normal bearing life of 4800 hours. The formula:

$$\text{New Bearing Life} = \frac{\text{Old Bearing Life} \times \text{Old Pump Speed}}{\text{New Pump Speed}}$$

Our example:

$$\text{New Life} = \frac{4800 \text{ Hours} \times 2000 \text{ rpm}}{4000 \text{ rpm}} = \frac{9,600,000}{4000}$$

$$= 2400 \text{ hours}$$

By doubling the speed, we have reduced the pump's life by one half. Now we can see why pump failures result from exceeding pump specifications.

CAVITATION

Pump cavitation is another "evil" resulting from either poor maintenance or poor operation.

Cavitation occurs when the fluid does not entirely fill the space provided for it in the pump. This leaves air or vapor cavities in the liquid which can be detrimental to the pump.

The combination of the high velocity (speed) of the discharged fluid and a restriction, usually caused by a clogged inlet line, between the reservoir and the pump causes the pressure of the incoming fluid to drop. When it is lowered, it cannot force enough fluid in to meet the demands of the pump. The result is that cavities or spaces are formed in the incoming fluid.

The pressure drops to the vapor tension of the fluid and the cavities fill with vapor. The vapor tension of the fluid is that pressure at which, at a given temperature, the fluid boils and freely evaporates. This evaporation fills the cavities.

The problem is further complicated by the pressure drop because this tends to release any dissolved air in the fluid and it, too, fills the cavities.

The damage to the pump results when these vapor-filled cavities, which have been formed in a low pressure area, meet with a high pressure area in the pump and are forced to collapse. This creates an action similar to an implosion which disintegrates or chips away small particles of the metal parts of the pump, adds excessive noise, and causes pump vibration.

A pump that is allowed to continually cavitate will soon have some seriously eroded parts, not to mention sluggish or erratic operation. Eventually the working parts may seize.

Chapter 1 covers the layout of hydraulic systems. Here we will only relate some of the pump problems that could develop if the system were changed in any way that would affect its operating efficiency.

Remember that the layout of each hydraulic system is precisely designed. The distance that the fluid must travel, the angle of the lines, the diameter of the lines, the placement of valves, filters and reservoirs—all are carefully planned to give the best performance of the overall system.

If a part of the system is changed and/or substandard parts are used, it could affect the operation of the pump and damage it.

For example, substituting a length of tubing that is smaller in diameter than the original, could create a pressure drop that would cavitate the pump.

Or placing the reservoir further from the pump could cause the pump to be starved for fluid.

A reservoir that is too small for the system will also cause starvation of the pump.

Using a line made of material that cannot meet the standards of the system and might burst under a surge of high pressure could also damage the pump.

POOR MAINTENANCE

Poor maintenance can also damage the pump. If hydraulic lines are not frequently checked for leaks, air could be introduced into the system, which could cause the fluid to foam or cavitate. Low fluid level in the reservoir could produce the same results. Failing to clean the system, when there is a chance that the fluid has become contaminated, is a prime way to damage the pump.

For details, see Chapter 11, "General Maintenance."

DIAGNOSING PUMP FAILURES

The following charts describe some of the ways in which the hydraulic pump can be damaged.

We have learned that much of the damage is due to carelessness or human error. To prevent these errors, follow the recommendations in the machine technical manual and follow the preventive maintenance program in the machine operator's manual.

The charts below should help you to remedy some of the problems that you may face in making a general diagnosis of hydraulic pump problems.

I. PUMP DOESN'T DELIVER FLUID

Possible Cause	Possible Remedy
1. Fluid level in reservoir too low.	1. Fill the reservoir with the proper grade and type of fluid. Check for possible external leaks.
2. Pump inlet line plugged.	2. Remove and clean. Check filters and reservoir for other possible obstructions.
3. Air leak in pump inlet line.	3. Repair leak.
4. Pump speed too slow.	4. Increase speed to within manufacturer's specifications. If belt-driven, check the belt and belt tension.
5. Sludge or dirt in the pump.	5. Dismantle and clean pump. Clean entire system and fill with clean fluid.
6. Fluid viscosity too high.	6. Check manufacturer's recommendation. Refill with same.
7. Variable control mechanism out of adjustment. (Variable displacement pumps)	7. Adjust according to the machine service manual specifications.
8. Broken or worn parts inside the pump.	8. Analyze the conditions that brought on the failure and correct them. Repair or replace the parts according to machine technical manual specifications.

II. NO PRESSURE

1. Pump not delivering fluid.	1. Follow the remedies in Part I above.
2. Vanes in vane pump sticking.	2. Check for burrs or metal particles that might hold vanes in their slots. Repair or replace if necessary. Clean system if contaminants are found.
3. Fluid recirculating back to reservoir and not going to functions.	3. Mechanical failure of some other part of the system, especially a relief valve. If contamination is involved, clean and refill with clean proper fluid.
4. Pump piston or valve broken or stuck open to allow fluid to return to inlet side.	4. Disassemble the pump, determine the cause, and correct it. Repair according to machine technical manual.

III. LOW OR ERRATIC PRESSURE

1. Cold fluid.	1. Warm up system. Operate only at recommended operating temperature range.
2. Fluid viscosity wrong.	2. Change to manufacturer's recommended grade.

III. LOW OR ERRATIC PRESSURE—Continued

3. Air leak or restriction at inlet line.

3. Repair or clean according to machine technical manual.

4. Pump speed too slow.

4. Increase speed to within specifications.

5. Internal parts in pump are sticking.

5. Dismantle and repair according to machine technical manual. Look for burrs on parts or metal particles in fluid. If contaminants are the the cause, thoroughly clean system and refill with proper grade of fluid.

6. Distance between internal parts has increased due to wear.

6. Dismantle and repair. If wear is abnormal, determine the cause by checking the operation and maintenance records as well as by examining the pump and system.

IV. PUMP MAKING NOISE

1. Restricted or clogged inlet line.

1. Clean or repair.

2. Air leaks in intake line or air drawn through inlet line.

2. Repair or make sure that inlet line is submerged in fluid in the reservoir. (To check for leaks, pour fluid around joints and listen for a change in sound of operation).

3. Low fluid level.

3. Refill with proper grade and type of fluid to the proper level.

4. Air in the system.

4. Check for leaks. Bleed air from the system.

5. Fluid viscosity too high.

5. Fill with fluid recommended by the manufacturer.

6. Pump speed too fast.

6. Operate pump within recommended speed.

7. Stuck pump part.

7. Check for solid contaminants in fluid or burrs on parts. If burrs are the cause, repair or replace part according to machine technical manual. If contaminated, thoroughly clean system and refill with proper fluid.

8. Worn or broken parts.

8. Check and correct cause of parts failure. Repair or replace bad parts.

V. EXCESSIVE WEAR

1. Abrasive contaminants or sludge in the fluid.

1. Check for cause of contaminants. Install or change fluid filter. Replace or repair worn parts according to machine service manual. Replace fluid with recommended grade and quantity.

2. Viscosity of fluid too low or too high.

2. Replace fluid with proper grade and type.

3. Sustained high pressure above maximum pump rating.

3. Check for possible relief valve malfunction or other parts failures.

4. Air leaks or restriction in system causing cavitation.

4. Eliminate from system. Check parts for degree of wear. Replace if necessary.

5. Drive shaft of pump misaligned.

5. Check and correct according to machine technical manual.

VI. EXCESSIVE FLUID LEAKAGE

1. Damaged seals or packings around drive shaft.

1. Check and replace. Check to be sure that chemicals in fluid are not destroying packing or seals. Follow manufacturer's recommendations on grade and type of fluid.

VII. INTERNAL PARTS BREAKAGE

1. Excessive pressure above maximum limits for pump.

1. Check for parts malfunction and cause. Repair according to machine technical manual.

2. Seizure due to lack of fluid.

2. Check reservoir fluid level, fluid inlet line for restriction, or plugged filter.

3. Abrasive contaminants in fluid getting by the filter.

3. See above.

TEST YOURSELF
QUESTIONS

1. (Fill in the blanks). A hydraulic pump converts _____ force into _____ force.

2. (True or false?) "Hydraulic pumps produce flow, not pressure."

3. (True or false?) "Non-positive displacement pumps are the best hydraulic pumps because they produce a continuous flow of fluid."

4. What three types of hydraulic pumps are generally used in modern farm and industrial hydraulic systems?

5. What is the major difference between an external gear pump and an internal gear pump?

6. A(n) _____ vane pump is capable of being constructed as a fixed or a variable displacement pump. (balanced-unbalanced)

7. What does "axial" and "radial" mean in reference to the piston pumps?

8. A _____ type pump usually has fewer moving parts than the other two types. (gear, vane, piston)

9. What is the most frequent cause of hydraulic pump failures?

10. What effect will doubling the operating pressure have on pump life?

11. Doubling pump speed will (increase-reduce) pump life by (¼ - ½ - ¾).

12. Why does a pump cavitate?

HYDRAULIC VALVES / CHAPTER 3

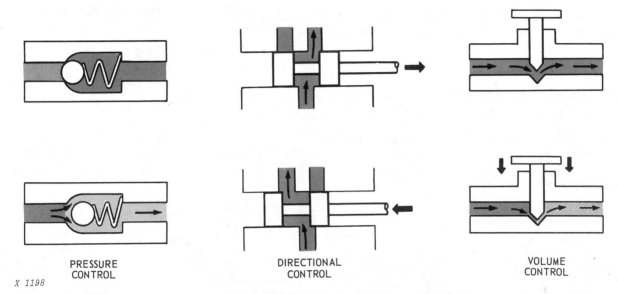

PRESSURE
CONTROL

DIRECTIONAL
CONTROL

VOLUME
CONTROL

X 1198

Fig. 1 — The Three Types of Valves

INTRODUCTION

Valves are the controls of the hydraulic system. They regulate the pressure, direction, and volume of oil flow in the hydraulic circuit.

Valves can be divided into three major types:

• **Pressure Control Valves**

• **Directional Control Valves**

• **Volume Control Valves**

Fig. 1 shows the basic operation of the three types of valves.

PRESSURE CONTROL VALVES are used to limit or reduce system pressure, unload a pump, or set the pressure at which oil enters a circuit. Pressure control valves include relief valves, pressure reducing valves, pressure sequence valves, and unloading valves.

DIRECTIONAL CONTROL VALVES control the direction of oil flow within a hydraulic system. They include check valves, spool valves, rotary valves, pilot controlled poppet valves, and electro-hydraulic valves.

VOLUME CONTROL VALVES regulate the volume of oil flow, usually by throttling or diverting it. They include compensated and non-compensated flow control valves and flow divider valves.

Some valves are variations of the three main types. For example, many *volume* control valves use a built-in *pressure* control valve.

Valves can be controlled in several ways: manually, hydraulically, electrically, or pneumatically. In some modern systems, the entire sequence of operation for a complex machine can be made automatic.

Let's discuss each type of valve in detail, starting with pressure control valves.

PRESSURE CONTROL VALVES

Pressure control valves are used to:

• **Limit system pressure**

• **Reduce pressures**

• **Set pressure at which oil enters a circuit**

• **Unload a pump**

RELIEF VALVES

Each hydraulic system is designed to operate in a certain pressure range. Higher pressures can damage the components or develop too great a force for the work to be done.

Relief valves remedy this danger. They are safety valves which release the excess oil when pressures get too high.

Two types of relief valves are used:

• DIRECT ACTING relief valves are simple open-closed valves.

• PILOT OPERATED relief valves have a "trigger" which controls the main relief valve.

Direct Acting Relief Valves

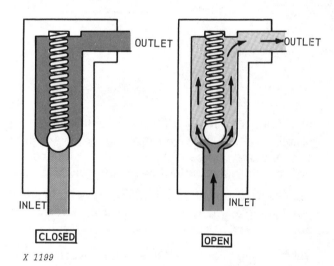

X 1199

Fig. 2—Direct Acting Relief Valve in Operation

Fig. 2 shows the operation of this simple valve. When closed, the spring tension is stronger than inlet oil pressure, holding the ball closed on its seat.

The valve opens when pressure rises at the oil inlet and overcomes the spring force. Oil then flows out to the reservoir, preventing any further rise in pressure.

The valve closes again when enough oil is released to drop pressure below the tension of the spring.

Some relief valves are adjustable. Often a screw is installed behind the spring (see Fig. 3). By turning the screw in or out, the relief valve can be adjusted to open at a certain pressure.

"Poppet" is a term for the working part of the valve. Ball poppets are most commonly used (though they may "chatter" during frequent operation). Other poppets used are shaped like buttons (Fig. 3) or like small cones or disks.

CRACKING PRESSURE AND PRESSURE OVERRIDE

"Cracking pressure" is the pressure at which the relief valve first begins to open. "Full-flow pressure" is the pressure at which the valve passes its full quantity of oil.

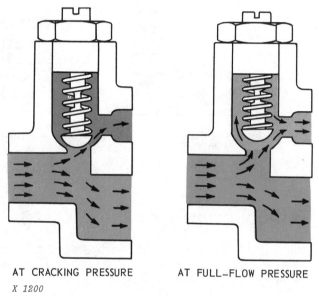

AT CRACKING PRESSURE   AT FULL-FLOW PRESSURE

X 1200

Fig. 3—Operation of Relief Valve Showing Pressure Override

Fig. 3 shows the oil flow during both of these cycles. Full-flow pressure is quite a bit higher than cracking pressure. This is because the spring tension builds up as the valve opens farther. This condition is called "pressure override" and it is one disadvantage of the simple relief valve.

USES OF DIRECT ACTING RELIEF VALVES

These valves are used mainly where volume is low, and for less frequent operations.

They have a fast response, making them ideal for relieving shock pressures. They are often used as safety valves to prevent damage to components.

Direct acting relief valves also serve as pilot valves for the pilot operated relief valves which are covered next in this chapter.

Direct acting relief valves are very simple. If they fail, no harm is usually done. The resulting pressure loss in the system is apparent to the operator and he can replace the broken spring or worn valve or seat.

Pilot Operated Relief Valves

When a relief valve is needed for large volumes with little pressure differential, a pilot operated relief valve is often used.

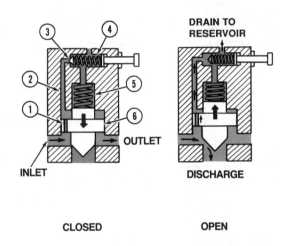

Fig. 4—Pilot Operated Relief Valve

The pilot valve is a "trigger" which controls the main relief valve. It is usually a small, spring-loaded relief valve built into the main relief valve (Fig. 4).

The main relief valve is closed when inlet oil pressure is below the valve setting. Passage (1) in the main valve (6) keeps it in hydraulic balance, while spring (5) holds it closed.

The pilot valve (3) is also closed at this time. Inlet pressure through sensing passage (2) is less than the pilot valve setting.

As inlet oil pressure rises, pressure in passage (2) also rises. When it reaches the pilot valve setting, the valve (3) is opened. This releases oil behind the main valve through passage (2) and out the drain port back to the reservoir. The resulting pressure drop behind the main relief valve (6) causes it to open. Now the main relief operation begins as excess oil is dumped at the discharge port, preventing a further rise in inlet pressure.

The valves close again when inlet oil pressure drops below the valve settings.

CRACKING PRESSURE AND PRESSURE OVERRIDE

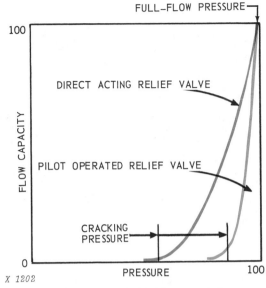

Fig. 5—Relief Valves Compared at Cracking and Full-Flow Pressures

Pilot operated relief valves have less pressure override than the simple direct acting types. Fig. 5 compares two of these valves. While the direct acting valve in Fig. 5 starts to open at about half its full-flow pressure, the pilot operated valve opens at about 90 percent of its full-flow pressure.

USES OF PILOT OPERATED RELIEF VALVES

Because these valves don't start to open until almost full-flow pressure, the efficiency of the system is protected—less oil is released. These valves are best for high-pressure, high-volume systems.

Although slower to operate than direct acting valves, the pilot operated relief valve keeps system oil at a more constant pressure while releasing oil.

PRESSURE REDUCING VALVES

A pressure reducing valve is used to keep the pressure in one branch of a circuit below that in the main circuit.

When not operating, a pressure reducing valve is *open* (Fig. 6). When it operates, it tends to close as shown.

Operation is as follows: When pressure starts to rise in the secondary circuit, force is exerted on the bottom of the valve spool, partly closing it. Spring tension holds the valve against the oil pressure so that only enough oil gets past the valve to serve

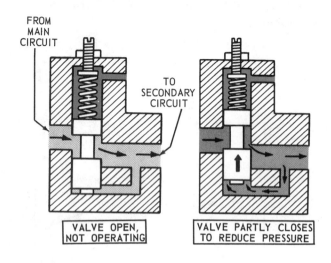

X 1203

Fig. 6—Pressure Reducing Valve

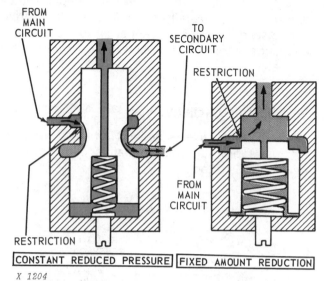

X 1204

Fig. 7—Two Types of Pressure Reducing Valves

the secondary circuit at the desired pressure. (The spring tension can be adjusted using the screw shown at the top.)

The pressure sensing for the valve comes from the *outlet* side, or the secondary circuit. This valve operates the reverse of a relief valve, which senses pressure from the inlet and is closed when not operating.

A pressure reducing valve will limit maximum pressures in the secondary circuit, regardless of pressure changes in the main circuit, as long as the system work load does not create back pressure into the reducing valve port. Back pressure would close the valve completely.

Two Types of Pressure Reducing Valves

Pressure reducing valves can operate in two ways:

• **Constant Reduced Pressure**

• **Fixed Amount Reduction**

CONSTANT REDUCED PRESSURE VALVES supply a fixed pressure regardless of main circuit pressure (so long as it is higher).

FIXED AMOUNT REDUCTION VALVES supply a fixed *amount* of pressure reduction, which means that it varies with the main circuit pressure. For example, the valve might be set to give a reduction of 500 psi. If system pressure was 2000 psi, the valve would reduce pressure to 1500 psi. If system pressure dropped to 1500 psi, the valve would reduce pressure to 1000 psi.

Operation of the two valves is shown in Fig. 7.

The *constant reduced pressure* valve operates by balancing the secondary pressure against an adjustable spring which is trying to open the valve. When secondary pressure drops, the spring opens the valve enough to increase pressure and to keep a constant reduced pressure in the secondary circuit.

The *fixed amount reduction* valve operates by balancing the main circuit pressure coming in against both the secondary pressure at the outlet and the spring pressure. Since the exposed areas of the inlet and outlet sides are equal, the fixed reduction will be that of the spring setting.

Note the restrictions shown in Fig. 7. These are the keys to pressure reduction. By partly closing, pressure reducing valves do their work.

For as we learned in Chapter 1, A RESTRICTION NORMALLY CAUSES A PRESSURE DROP.

Pilot Operated Types

As with relief valves, a small pilot valve can be added to control a pressure reducing valve.

Operation is the same as described above except that the pilot valve acts first to "trigger" the reducing valve.

Using a pilot valve gives a wider range of adjustment and a more consistent relief operation.

PRESSURE SEQUENCE VALVES

Pressure sequence valves are used to control the sequence of flow to various branches of a circuit. Usually the valves allow flow to a second function only after a first has been fully satisfied.

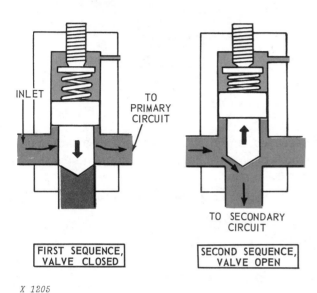

X 1205

Fig. 8—Pressure Sequence Valve in Operation

Fig. 8 shows a pressure sequence valve in operation.

When closed, the valve directs oil freely to the primary circuit.

When opened, the valve diverts oil to a secondary circuit.

The valve opens when pressure oil to the primary reaches a preset point (adjustable at the valve spring). The valve is then lifted off its seat as shown and oil can flow through the lower port to the secondary.

One use of the sequence valve is to regulate the operating sequence of two separate cylinders. The second cylinder begins its stroke when the first completes its stroke. Here the sequence valve keeps pressure on the first cylinder during the operation of the second.

Sequence valves sometimes have check valves which allow a reverse free flow from the secondary to the primary, but sequencing action is provided only when the flow is from primary to secondary.

UNLOADING VALVES

The unloading valve directs pump output oil back to reservoir at low pressure after system pressure has been reached. They may be installed in the pump outlet line with a tee connection.

In some hydraulic systems, pump flow may not be needed during part of the cycle. If pump output has to flow through a relief valve at system pressure, much hydraulic energy is wasted as heat. This is where an unloading valve works best.

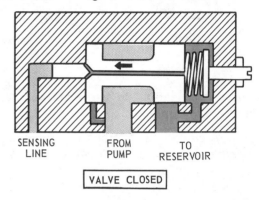

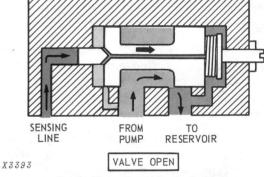

X3393

Fig. 9—Unloading Valve in Operation

When closed (Fig. 9), spring pressure holds the valve on its seat. Sensing pressure at the other end of the valve is less than spring pressure. The reservoir outlet is closed and no unloading occurs.

The valve opens when the sensing pressure rises and overcomes the spring thrust. The valve moves back, opening the outlet to the reservoir. Pump output oil is now diverted to the reservoir at low pressure.

Unloading Valves for Accumulator Circuits

An unloading valve is often used in an accumulator circuit to unload the pump after the accumulator is charged. (See Chapter 6 for details on accumulators.)

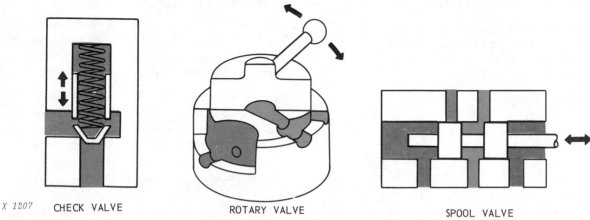

X 1207 CHECK VALVE   ROTARY VALVE   SPOOL VALVE

Fig. 10—The Three Types of Directional Control Valves

The valve is closed while the pump charges the accumulator with oil.

As the pressure rises it forces the small sensing piston against the large valve and compresses the spring. When accumulator pressure reaches that determined by the spring setting, the valve opens by passing oil and relieving the pump.

At this time the low neutral pressure oil is directed to the large end of the large piston.

When the accumulator discharges and the system pressure drops, the spring moved the valve against the reduced system pressure in the small piston and the neutral pressure against the large end of the valve.

This means the valve will close at a slightly lower pressure than it opens. This gives the valve an operating range and prevents chattering.

DIRECTIONAL CONTROL VALVES

Directional control valves direct the flow of oil in hydraulic systems. They include these types:

- **Check Valves**
- **Rotary Valves**
- **Spool Valves**
- **Pilot Controlled Poppet Valves**
- **Electro-Hydraulic Valves**

The first three types of directional control valves are compared in Fig. 10. Each uses a different type of valving element to direct oil. The *check valve* uses a poppet which seats and unseats; the *rotary valve* uses a rotary spool which turns to open and close oil

passages; and the *spool valve* uses a sliding spool which moves back and forth to open and close oil routes.

Let's discuss each valve in more detail.

CHECK VALVES

Check valves are simple one-way valves. They open to allow flow in one direction, but close to prevent flow in the opposite direction.

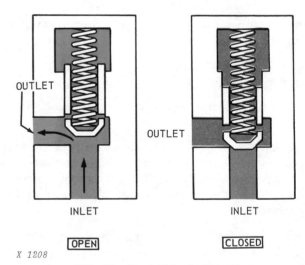

OUTLET

OUTLET

INLET

INLET

OPEN

CLOSED

X 1208

Fig. 11—Check Valve In Operation

Fig. 11 shows a simple check valve in operation.

The valve is opened by system pressure which pushes the valve up against its spring. Oil then flows freely past the valve as shown.

The valve closes when inlet pressure drops. This stops reverse flow and traps pressure oil already in the circuit.

Check valves are usually installed in an oil line. They may also be part of some other valve, such as a sequence or pressure reducing one.

Although a check valve is most often used to stop reverse flow, sometimes reverse flow is needed during one phase of circuit operation. In this case, pilot operated check valves have been devised. For example, a check valve might be used in a cylinder line to prevent leakage, but an extra function is needed to allow reverse flow when the cylinder must be stroked. A pilot piston which can force the check valve open during these cylinder strokes will handle the job.

ROTARY DIRECTIONAL VALVES

Rotary valves are commonly used as pilot valves to direct flow to other valves.

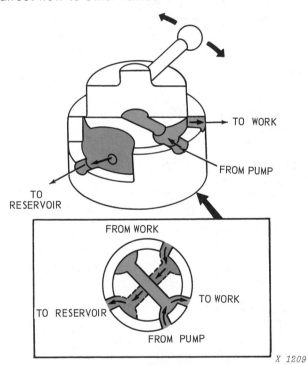

Fig. 12—Rotary Directional Valve

Fig. 12 shows a four-way rotary valve. The valve has holes which match with holes in the main body as the valve turns. The valve is turned by a hand lever; other models may be operated hydraulically or electrically.

Fig. 12 shows the valve positioned to allow pressure oil from pump to enter one port, flow through the valve, and out another port to the work. Meanwhile, oil is returning from another work port through the valve to reservoir. The drilled ports in the valve are actually on two levels to separate them.

Rotary valves can be modified to operate as two-, three-, or four-way valves. This is done by relocating the ports, altering the passages, or adding and removing oil routes. Two-way models are simple on-off shutoff valves; three- and four-way models are usually designed as pilot valves.

Rotary valves are generally used as low-pressure, low-volume controls. They are simple and compact enough for use as pilots for other valves in more complex systems.

SPOOL DIRECTIONAL VALVES

The sliding spool valve is a true directional control. Used as a "control valve," it directs oil to start, operate, and stop the actuating units on most modern hydraulic systems.

There is no limit to the variations in spool valve design. Two-, four-, and six-land spools are most common, often used in valve "stacks" of two or more. In this case, each spool controls a branch of the circuit.

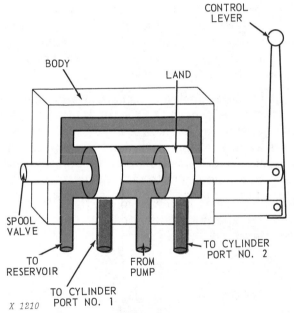

Fig. 13—Spool Directional Valve

Fig. 13 shows a simple two-land spool valve. Moving the spool from neutral (shown) to the right or left opens up some passages and closes others. In this way it directs oil to and from the actuating cylinder. The spool lands seal off the inlet from the outlet oil.

The spool is usually hardened and ground to produce a smooth, accurate, and durable surface. It may also be chrome-plated to resist wear, rust, and corrosion.

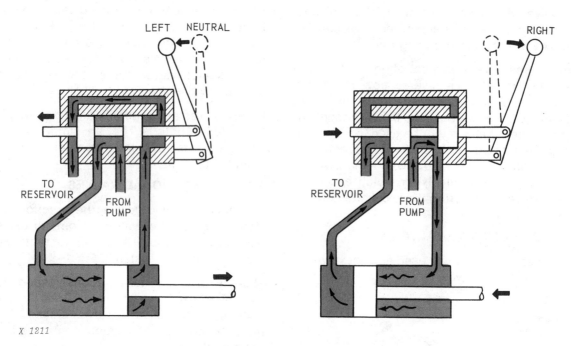

Fig. 14—Spool Valve Directing Oil to Cylinder

The spool valve shown in Fig. 13 is called a "three-position, four-way" valve. The valve has three positions: neutral, left, and right. And it is connected to the circuit in four ways: to pump, to reservoir, to cylinder port 1, and to cylinder port 2.

Fig. 14 shows the same spool valve in operation. As the valve is moved to the left, it directs oil from the pump to the left side of the cylinder, actuating it as shown. At the same time, the valve opens a passage which allows oil from the opposite end of the cylinder to return to the reservoir.

When the valve is moved to the right, the flow is reversed and the cylinder operates in the opposite direction.

In neutral (see Fig. 13), the spool valve lands seal off both cylinder ports, trapping oil to hold the cylinder in place.

Open and Closed Center Spool Valves

In Chapter 1 we covered the two kinds of hydraulic systems, open and closed center. Each uses a different type of spool valve (Fig. 15).

• OPEN CENTER spool valves allow pump oil to flow *through* the valve during neutral and return to reservoir.

• CLOSED CENTER spool valves *stop* (deadend) the flow of oil from the pump during neutral.

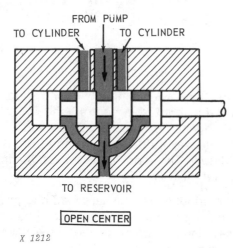

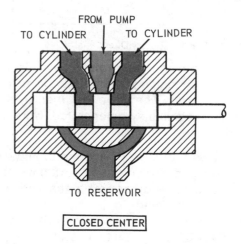

Fig. 15—Open and Closed Center Spool Valves (In Neutral)

Normally the cylinder ports are blocked when a spool valve is in neutral. However, in some designs the ports are open to allow the cylinder to "float."

Control of Spool Valves

Spool valves can be controlled manually or they may use pilot valves, electrical solenoids, or hydraulic oil acting on the ends of the spool.

Detent mechanisms are sometimes used to hold the valve in position during each operation.

Multiple Uses of Spool Valves

Two or more spool valves can be used in one compact control package to operate several functions. This can be done in two ways:

• VALVE STACKS—several sections of valves bolted together.

• "UNIBODY" VALVES—several valves in one solid housing.

Fig. 16 shows the two kinds of spool valve packages.

Valve stacks allow extra valves to be added easily by inserting another section in the package. However, care must be taken in sealing the mating surfaces of each section.

"Unibody" valves are less flexible but more permanent. Oil leakage is less of a problem since one solid housing holds all the valves. However, if one valve bore is damaged it may mean replacing the whole valve housing.

Both valve packages normally use one common oil inlet and outlet to the system. Either package can be designed for open or closed center operation.

General Uses of Spool Valves

Spool valves are popular on modern hydraulic systems for several reasons:

1) *Quick, positive action.* Spool valves can be precision ground for fine oil metering.

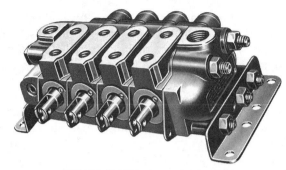

X 1234   VALVE STACKS

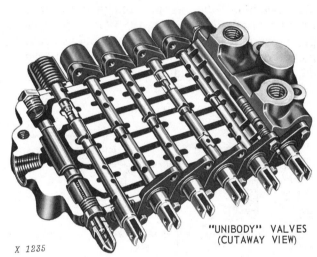

X 1235

"UNIBODY" VALVES
(CUTAWAY VIEW)

Fig. 16—Multiple Uses of Spool Valves

2) *Adaptibility.* By adding extra lands and oil ports, spool valves can be made to handle flows in many directions.

3) *Compactness.* Stacking of spool valves in one compact control package is easy. This is very important on mobile systems.

However, spool valves also require good maintenance. Dirty oil will damage the mating surfaces of the valve lands, causing them to lose their accuracy. Dirt will also cause these valves to stick or work erratically.

Above all, spool valves must be accurately machined and fitted to their bores.

PILOT-CONTROLLED POPPET VALVES

This type of valve is used in areas where remote mounting of the valve is desirable. The valve may be mounted close to the function it controls. This eliminates the need for routing hydraulic pipes and hoses over long distances for every function.

The valve reduces problems due to valve leakage (such as cylinder drift) because it is a low-leakage valve. Each valve is flow-adjustable to vary the amount of oil flow.

Pilot controlled poppet valves (Fig. 17) control the flow of pressure oil to a function and return oil from that function. Thus two poppets (a pressure and return poppet) are used (Fig. 18). When the pressure poppet opens, the return poppet also opens because a shaft connects the two (Fig. 18, right). This allows oil to flow to and from a function.

The poppet valve is actuated by a pilot valve. A pressure differential is created on the bottom side of the pressure poppet. The flow of oil required to create this differential is controlled by a pilot valve (Fig. 19). The pilot valve is opened either manually or by an electric solenoid.

When the pilot valve is opened, pilot oil is allowed to escape from the cavity below the pressure poppet (Fig. 19, right). The pressure oil coming into the valve body housing is able to push the pressure poppet down, thus opening the valve and causing the function to move. The valve will remain open as long as the pilot valve is open.

When the pilot valve is closed (left, Fig. 19), the pilot oil can no longer escape. Oil from the pressure oil passage refills the pilot oil area under the pressure poppet through a groove on the outside edge of the poppet (Fig. 19). The pilot oil pressure increases to the same pressure as the valve inlet oil. Remember, the area of the bottom of the pressure poppet is larger than the area on top of the poppet.

Consequently, with the oil pressure equal on both sides of the poppet, the pressure poppet is pushed up against its seat to close the valve and stop oil flow to the function.

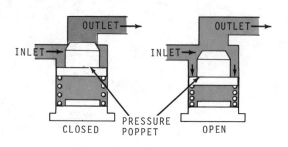

X7605

Fig. 17 — Basic Poppet Valves (Closed and Open)

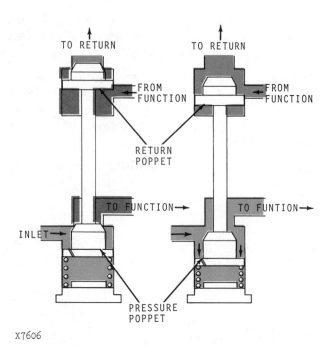

X7606

Fig. 18 — Poppet Valves - Pressure and Return

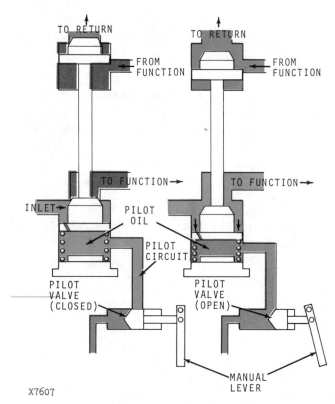

X7607

Fig. 19 — Pilot Control of Poppet Valves

Multiple Poppet Valve Assemblies

The preceding discussion shows how a single poppet valve controls the flow of pressure oil in one direction. Therefore, to control a function in two directions, two poppet valve assemblies are needed.

A cylinder is connected to the control valve work ports (Fig. 20) — the rod end of the cylinder to work port A and the head end of the cylinder to work port B.

When the pilot valve for poppet assembly A opens (Fig. 20), pressure oil pushes the pressure poppet for assembly A down. The return poppet opens at the same time as the pressure poppet.

Pressure oil then flows into a passage connecting the cavity above pressure poppet A to work port A. The oil flows to the rod end of the cylinder, causing the cylinder to retract. The connecting passage allows pressure oil to flow into and through the opening where return poppet of assembly B is located. With the B return poppet closed pressure oil is then directed to work port A.

Return oil from the cylinder flows into work port B, through the cavity above the pressure poppet for assembly B, to a second oil passage. This passage directs the return oil from work port B to the open return poppet A. The return oil then flows past the open return poppet into the return oil cavity and eventually into the return oil circuit.

Poppet valve assembly B controls the flow of pressure oil out of the work port B (Fig. 21). Again, opening the pilot valve for the poppet assembly B creates a pressure drop under the pressure poppet for assembly B. System pressure oil then forces the pressure poppet open and at the same time opens return poppet for assembly B.

Pressure oil flows past the opened pressure poppet, to the work port B, and to the head end of the cylinder. The cylinder extends.

Return oil from the cylinder enters the work port A and flows past the return poppet B into the return oil cavity which is connected to the return circuit.

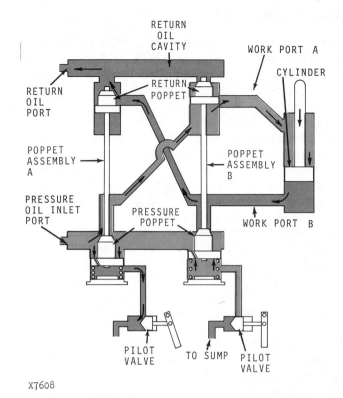

X7608

Fig. 20 — Poppet Valves Retracting Cylinder

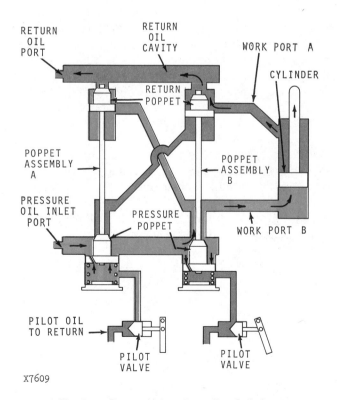

X7609

Fig. 21 — Poppet Valves Extending Cylinder

ELECTRO-HYDRAULIC VALVES

Hydraulic control valves may be actuated by an electric solenoid. Solenoids are designed to do mechanical jobs by means of electromagnets (see FOS Electrical Systems).

Solenoid-controlled valves are used in places where the valves are located near the functions they control. In these cases, solenoid-controlled valves eliminate the need for long hydraulic hoses and pipes to be routed to each function. The valves are controlled by the electrical system.

Solenoids used to control the flow of hydraulic oil basically consist of contacts and windings around a hollow cylinder. The cylinder contains a movable core or valve. When the winding is energized by battery current, the solenoid valve is pulled upward, giving the necessary mechanical movement.

When the rocker switch is pressed down on the left side (Fig. 22), the solenoid on the left becomes energized. The solenoid valve retracts upward, allowing pressure oil from the main hydraulic pump to enter the hollow center of the solenoid valve. Oil is directed through the solenoid valve to the left side of the direction valve.

The solenoid on the right is not energized and spring pressure forces the solenoid valve down. This closes the passage and restricts pressure oil from entering the hollow center of the right solenoid valve. Now pressure free oil is allowed to flow from the right side of the direction valve, through the right solenoid valve, and back to the reservoir.

Due to the pressure differential, the direction valve moves to the right, opening the passage for pressure oil to the left side of the cylinder.

Pressure free oil from the right side of the cylinder flows through the direction valve and back to the reservoir.

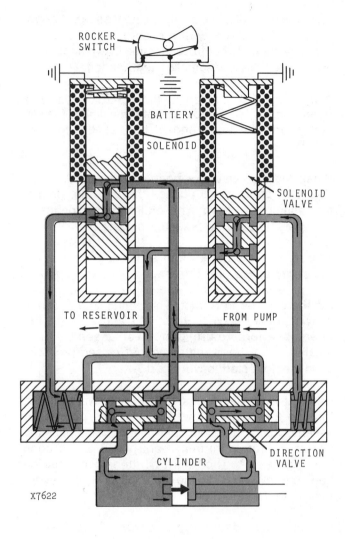

Fig. 22 — Left Solenoid Energized

When the rocker switch is placed in the neutral position (Fig. 23), both solenoid valves assume a relieved, downward position. This restricts pressure oil from entering either solenoid valve. Equalized spring pressure then centers the direction valve, cutting off pressure oil flow to the cylinder. Oil is trapped on both sides of the cylinder.

When the rocker switch is pressed down on the right side, the right solenoid becomes energized (Fig. 24). The action of the valves and oil flow reverses the circuit from when the rocker switch was energizing the left solenoid.

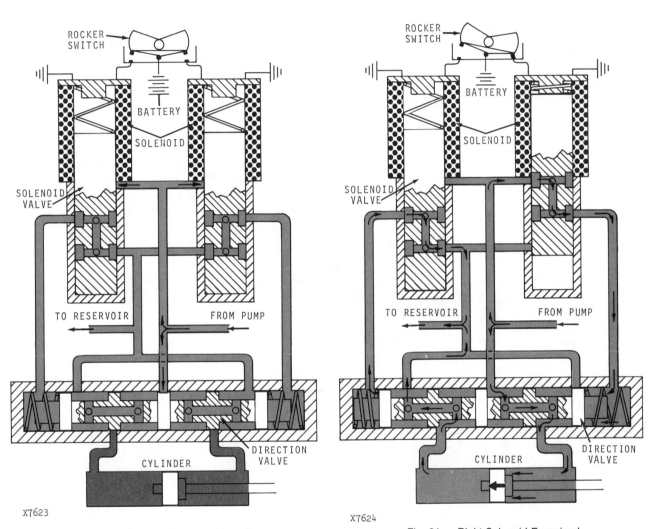

Fig. 23 — Rocker Switch in Neutral   Fig. 24 — Right Solenoid Energized

MICROPROCESSOR-BASED ELECTRONIC CONTROL OF HYDRAULIC VALVES

On pages 12 and 13 of this chapter we discussed the function of electro-hydraulic valves. Control of these valves can be taken over by a microprocessor, and the required sensors and circuitry. There are a number of advantages, which can be gained when a hydro-mechanical system is replaced by a microprocessor-based electronic system for the control of the valves. Here are some of them:

- **Elimination of mechanical linkages**
- **Flexibility in design**
- **Controller can be "taught"**
- **Greater accuracy of control**

Application

A typical application of a microprocessor-based electronic system for the control of electro-hydraulic valves is the electro-hydraulic hitch on newer farm tractors. (Fig. 25).

Let's look at the function. Draft is sensed by strap (1), which uses a strain gauge sensor. The hitch position is sensed by potentiometer (3) attached to one of the rockshaft lift arms. The outputs of these sensors are read directly by hitch controller (2) and are summed by the hitch algorithm, which uses the setting of mix potentiometer (4) to determine the relative weight of each. The rockshaft control lever is attached to rotary potentiometer (5), which provides the lever command to the algorithm. Rate-of-drop potentiometer (7) and raise limit potentiometer (8) have also been added.

The lever command, the mix setting and the draft and position feedbacks are used to determine the command to send to the appropriate (pressure or return) valve solenoids (6). The electro-hydraulic hitch control algorithm is defined by the engineer in the design stage, using predetermined equations. Fig. 26 shows the schematic circuitry.

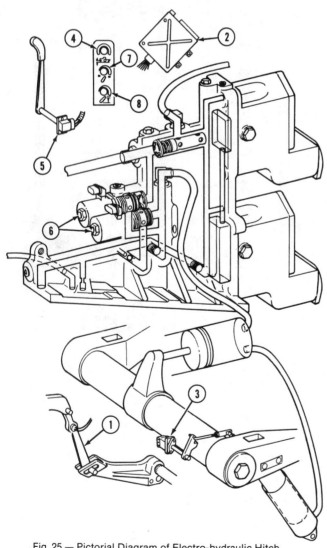

Fig. 25 — Pictorial Diagram of Electro-hydraulic Hitch

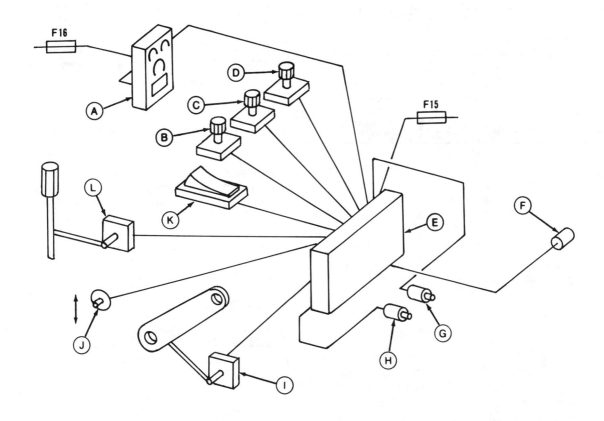

Fig. 26 — Electro-hydraulic Hitch Circuit

A — Tachometer Module
B — Raise Limit Pot
C — Rate-of-Drop Pot
D — Load/Depth Pot

E — Hitch Control Unit (HCU)
F — Load/Draft Sensor
G — Pressure Solenoid Valve
H — Return Solenoid Valve

I — Rockshaft Position
　　Feedback Sensor
J — External Raise/Lower
　　Switch

K — Raise/Lower Rocker
　　Switch
L — Hitch Control Lever
　　Position Sensor

The electro-hydraulic hitch uses electrical circuitry and solenoids to control hydraulic oil flow which activates the 3-point hitch.

The electrical portion of the hitch consists of five major groups:

1. **Hitch Control Unit (HCU)**
2. **Operator Controls**
   - **Hitch control lever**
   - **Load/depth control pot**
   - **Rate-of-drop pot**
   - **Raise limit pot**
   - **Raise/lower rocker switch**
   - **External raise/lower switch**

3. **Hitch sensing devices**
   - **Rockshaft feedback position sensor**
   - **Load/draft sensor**
4. **Control Valves**
   - **Pressure and return valve solenoids**
5. **Wiring Harnesses**

VOLUME CONTROL VALVES

Volume control valves control the volume or flow of oil, usually by throttling or diverting it (Fig. 27)

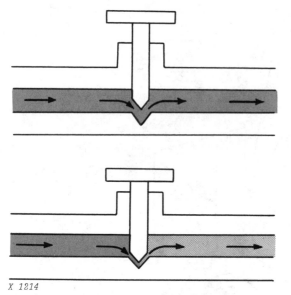

X 1214

Fig. 27 — Basic Principle of Volume Control Valve

In many hydraulic systems, the speed of a cylinder or motor must be closely regulated. This can often be done by regulating the *volume* of oil flowing to the actuator. When using a fixed displacement pump, the normal way to regulate the speed of a cylinder or motor is with volume control valves.

Volume control valves can be separated in two types:

• FLOW CONTROL VALVES which control volume flow, usually through a metering orifice.

• FLOW DIVIDER VALVES which control volume flow but also divide the flow between two or more circuits.

Let's cover each type in more detail.

FLOW CONTROL VALVES

Flow control valves can regulate flow by:

1) Restricting flow in or out of the component whose speed is being regulated. These valves are *noncompensated.*

2) Diverting flow away from the component whose speed is being regulated. These valves are usually *compensated.*

Noncompensated valves do not compensate for pressure changes. As inlet flow changes, so does the flow through the valve. These valves are not generally used where accurate flow rate control is needed. They include simple needle and globe valves.

Compensated valves maintain a constant flow even though the inlet flow changes. These valves adjust the flow to make up for raises and drops in the inlet flow.

Needle And Globe Valves (Noncompensated)

These "noncompensated" valves are used in many hydraulic circuits. While not sensitive to pressure changes, they are simple and can be adjusted to meter oil flow very carefully and exactly.

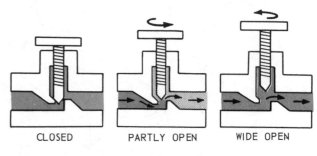

CLOSED   PARTLY OPEN   WIDE OPEN

X 1215

Fig. 28 — Needle Valve

The needle valve (Fig. 28) is a simple restrictor. When the pointed stem is screwed down into the orifice, flow is shut off. When unscrewed a small amount, flow is limited (and pressure drops). When screwed out farther, the stem allows full flow.

A common use for the needle valve is in a tractor rockshaft. Here it regulates the speed of drop of the rockshaft and mounted tool.

The globe valve operates the same except it has a rounded metering tip. (Often globe valves are used in water plumbing systems where finer metering is not needed.)

Compensated Flow Control Valves

This valve operates on the principle that with a given sized orifice and with a controlled pressure drop across that orifice, the flow will remain constant (see Fig. 29, top). The fixed orifice on the end of the spool is matched to the spring.

When more than the predetermined flow tries to go through the orifice, the pressure difference be-

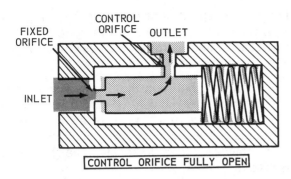

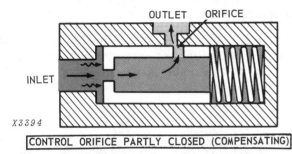

Fig. 29 — Flow Control Valve Which is Pressure Compensated

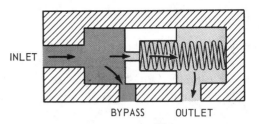

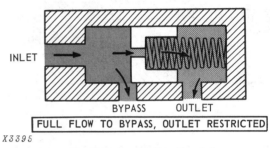

Fig. 30 — Bypass Flow Regulator

tween the front and inside of the valve increases. This compresses the spring and moves the valve to restrict the flow at the outlet orifice. This increases the pressure on the inside of the valve and reduces flow through the fixed orifice.

Regardless of what changes are made in the pressure of the working circuit or inlet pressure, the spring will maintain the same pressure drop and, therefore, the same flow through the fixed orifice.

This valve is used in closed-center systems where flow variations are controlled by the pump.

Bypass Flow Regulation

Another form of flow control valve is the bypass flow regulator.

This valve is used in open center systems, when the total output of the pump must either be used in the function, passed on to another function, or returned to the reservoir. The regulator valve also uses the principle of the spring and fixed orifice to control flow.

The pressure on the inside of the valve is determined by the working pressure of the function (outlet). (See Fig. 30, top.) The valve then maintains the inlet pressure enough higher to maintain the proper differential across the orifice. This is done by dumping or bypassing excess oil. As flow increases, pressure builds on the head of the valve. This pushes the valve back, enlarging the bypass

opening and maintaining the same pressure as before (Fig. 30, bottom).

Bypass oil can be directed to another function or back to the reservoir. When directed to another function, the valve becomes a priority valve, insuring that the first predetermined amount of oil goes to a primary function outlet passage and the balance to a secondary function (bypass).

A relief valve is installed in the outlet of this valve to protect the system against excess pressures from "surges," which can close the bypass port completely.

FLOW DIVIDER VALVES

Flow divider valves control volume flow but also divide the flow between two or more circuits. They do this in three ways:

• PRIORITY valves deliver all fluid to one circuit until pump delivery exceeds the needs of that circuit; then the extra delivery is made available to other circuits.

• ADJUSTABLE PRIORITY valves do the same as above except that delivery to the No. 1 (priority) circuit can be adjusted.

• PROPORTIONAL valves deliver flow to all circuits at all times. However, the delivery to each circuit can be modified. For example, with two circuits the proportion can be 50-50 or up to a 90-10 ratio.

Priority Flow Dividers

The priority flow divider shown in Fig. 31 is built into a hydraulic pump. It divides the pump flow into two separate outlets. One outlet has priority; the other receives oil after the first is satisfied.

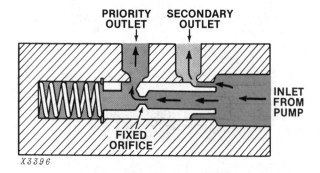

Fig. 31 — Priority Flow Divider

The flow divider spool slides in its bore, opening one outlet wider as it restricts the other. Inlet oil presses against one end of the valve spool, while a spring and lower pressure oil push on the other end.

When pump output is lower, the valve moves to the right, opening the priority outlet wider. An orifice in the valve meters oil for this outlet.

As pump output rises, the pressure drop through the valve orifice increases and causes the valve to compress the spring, moving to the left. This opens the secondary port wider and feeds it the excess oil. The priority port still gets its oil, but the secondary gets all the remaining oil.

As an example, the priority port might serve the power steering of a machine, while the secondary feeds a loader circuit. The loader needs more oil, but the steering is more vital to operation of the machine. Let's say the pump has an output of 10 gpm at full speed, while the power steering needs only 2 gpm. At full speed the steering gets 2 gpm while the loader gets 8 gpm. At low speed the pump may produce only 2 gpm. In this case the priority outlet for the steering gets all the pump's flow. At intermediate speeds the flow is divided at different ratios, but the steering always gets its 2 gpm first.

In adjustable priority valves, the delivery to the priority port can be adjusted using external levers, solenoids, or hydraulic balancers, or internally, by changing valve spring tension or by shimming the valve. A relief valve is used with the priority valve to protect it against surge pressures which might close off the secondary outlet.

Note the similarity between the priority flow divider and the bypass flow regulator (Fig. 30). Operation is similar, but the result is different: the flow divider feeds two working circuits while the bypass regulator only feeds one, exhausting the remaining oil back to reservoir.

Proportional Flow Divider

This valve simply takes a single flow of oil and divides it between two circuits (Fig. 32). The dividing of oil may be 50-50 or in ratios up to 90-10.

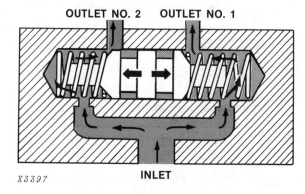

Fig. 32 — Proportional Flow Divider

The flow divider valve shown is one used on a crawler power steering system. It sends an equal amount of oil to the left and right steering valves. This is done by having the two orifices from the inlet line to the ends of the spool equal in size.

When the right steering valve is actuated, the pressure backup from the valve will move the spool to the left. It will restrict the left port enough to maintain equal pressure (pressure required to steer) on each side of spool. The spool is free floating so this balance will always be maintained.

Because the inlet pressure to each orifice is equal, because the pressure at each end of the spool is equal, and because the flow through an orifice is the same with the same pressure drop, there will always be equal flow to each steering valve, regardless of which valve is used.

To divide the flow on other than a 50-50 basis, it is necessary only to vary the proportional size of the two orifices.

MISCELLANEOUS VALVES
AUTOMATIC AIR VENT VALVES

The automatic air vent valve is used to keep the hydraulic system free from air.

The valve is placed at the highest point in the system where air will collect. Liquid pressure in the system and in the valve keeps it closed.

As air accumulates in the body of the valve, it displaces the liquid, and as the liquid recedes, the valve opens. The liquid, being under pressure, forces out the air and "bleeds" the system. As the air leaves, the liquid in the valve body rises and closes the valve, preventing the escape of oil. When air collects again, the valve repeats the cycle.

GATE VALVES

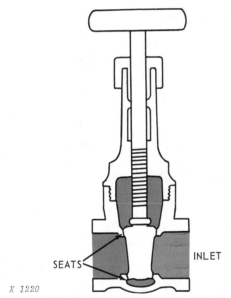

X 1220

Fig. 33 — Gate Valve in Closed Position

A gate valve is used to open or close a line to flow. The valving element is a wedge-shaped gate which is raised or lowered by screw action (Fig. 33). This valve is designed to completely open or close a line, but not to throttle flow when partly open.

Although a gate valve offers very little resistance to flow when completely open, it is difficult to open and close under high pressure.

COCK VALVES

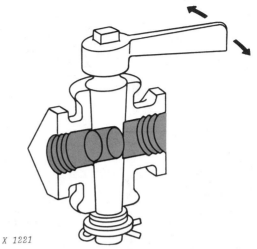

X 1221

Fig. 34 — Cock Valve in Open Position

Cock valves are very simple and are usually small in size. They are used to bleed air out of a system, turn gauges on and off, or drain the system. Fig. 34 shows a cock valve in the open position. A quarter-turn of the handle will shut it off. The valve shown is designed for moderate pressures. Modifying the valves will make it work for much higher pressures.

FLAPPER VALVES

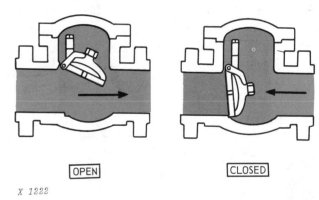

X 1222

Fig. 35 — Flapper Valve

Flapper valves are essentially check valves. They permit flow in only one direction. They are made in all sizes—very small to very large. They offer little resistance to flow when fully open. Although they are usually installed so that gravity and pressure close them, they sometimes have a spring to start closing the flapper. Back pressure causes the flapper valve to seal tightly.

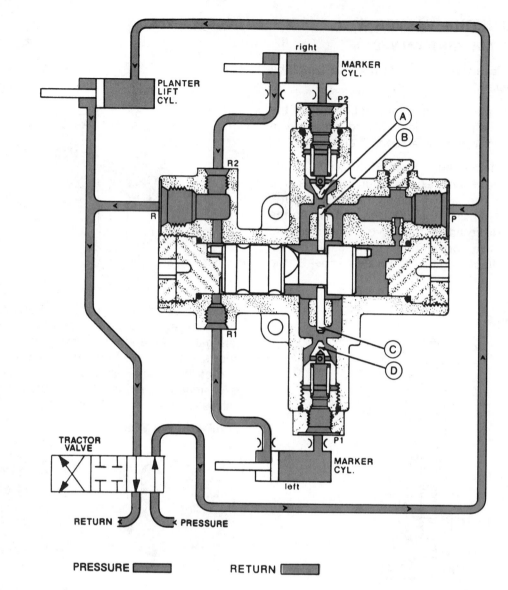

A30873

PRESSURE ▭   RETURN ▭

Fig. 36 — A Typical Sequence Valve

SEQUENCE VALVE

A sequence valve is used to control more than one movement in a machine in sequence. For example, the logic behind the application of a sequence valve is that **if** action A occurs, **then** action B will take place. This concept is applied to machines such as corn planters. If the lift lever is actuated for the planter, then the planter markers will automatically raise.

Fig. 36 shows the hydraulic fluid flow when a planter is raised with a sequence valve. In the first step, the valve spool actuates the planter lift cylinder. Second, poppets A and D if both markers are down, are opened. Oil flows to the rod side of the cylinder and raises the down marker or markers. Pin C and B control the downward movement of the markers in this specific example by blocking oil flow to the lift side or rod end of the cylinder.

VALVE FAILURES AND REMEDIES

INTRODUCTION

Hydraulic valves are precision-made and must be very accurate in controlling the pressure, direction, and volume of fluid within a system. Generally, no packings are used on valves since leakage is slight as long as the valves are carefully fitted and kept in good condition.

Contaminants such as dirt in the oil are the major villains in valve failures. Small amounts of dirt, lint, rust or sludge can cause annoying malfunctions and extensively damage valve parts. Such material will cause the valve to stick, plug small openings, or abrade the mating surfaces until the valve leaks. Any of these conditions will result in poor machine operation, or even complete stoppage. This damage may be eliminated if operators use care in keeping out dirt.

Use only the specified oils in the hydraulic system. Follow the recommendations in the machine operator's manual. Because oxidation produces rust particles, an oil that will not oxidize must be used. The oil should be changed and the filters serviced regularly.

For successful valve service, observe the following precautions.

BEFORE SERVICING VALVES

1. Disconnect electrical power source before removing hydraulic valve components to eliminate accidental starting or tools shorting out.

2. Move valve control lever in all directions to release hydraulic pressure in the system before disconnecting any hydraulic valve components.

3. Block up or lower all hydraulic working units to the ground before disconnecting any parts.

4. Clean the valve and surrounding area before removing any part for service. Use steam cleaning equipment if available; however, DO NOT ALLOW WATER TO ENTER THE SYSTEM. Be certain all hose and line connections are tight.

5. If steam cleaning is not possible, use fuel oil or other suitable solvent. Never use paint thinner or acetone as a cleaning agent. Plug port holes immediately after disconnecting lines.

VALVE DISASSEMBLY HINTS

1. Do not perform hydraulic valve internal service work on the shop floor, on the ground, or where there is danger of dust or dirt being blown into parts. USE ONLY A CLEAN BENCH AREA. Be certain all tools are clean and free of grease and dirt.

2. During disassembly, be careful to identify the parts for reassembly. Spools are *selectively fitted* to valve bodies and must be returned to the *same* bodies from which they were removed. Valve sections must be reassembled in the same order.

3. When it is necessary to clamp a valve housing in the vise, use extreme caution. Do not damage the component. If possible, use a vise equipped with lead or brass jaws or protect the component by wrapping it in a protective covering.

4. All valve housing openings should be sealed when components are removed during service work. This will prevent foreign material from entering the housing.

 CAUTION: On spring-loaded valves, be very careful when removing the back-up plug as personal injury may result.

5. When springs are under high preload, use a press to remove them.

6. Wash all valve components in a clean mineral oil solvent (or other non-corrosive cleaner). Dry parts with compressed air and place on a clean surface for inspection. *Do not wipe valves with waste paper or rags.* Lint deposited on any parts may enter the hydraulic system and cause trouble.

7. DO NOT USE CARBON TETRACHLORIDE as a cleaning solvent as it causes deterioration of rubber seals.

8. After parts are cleaned and dried, coat them immediately with a rust-inhibiting hydraulic oil. Then be sure to keep the parts clean and free of moisture until they are installed.

9. Carefully inspect valve springs during valve disassembly. Replace all springs that show signs of being cocked or crooked, or contain broken, fractured or rusty coils.

Use a spring tester to check strength of springs in pounds, compressed to a specified length (Fig. 37).

VALVE REPAIR

Directional Control Valve Repair

Directional control valve spools are installed in the valve housing by a select hone fit. This is done to provide the closest possible fit between housing and spool for minimum internal leakage and maximum holding qualities.

T 9600

Fig. 37 — Checking Valve Spring Tension

To make this close fit, special factory techniques and equipment are required. Therefore, most valve spools and bodies are furnished for service only in MATCHED SETS and are not available individually for replacement.

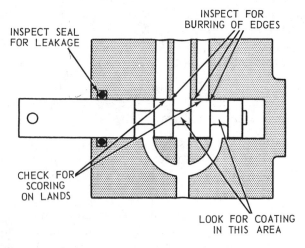

Fig. 38 — Inspecting Directional Control Valve Spools and Bores

When repairing, inspect the valve spools and bores for burrs and scoring in the areas shown in Fig. 38. Spools may become coated with impurities from the hydraulic oil. When scoring or coating is not deep enough to cause objectionable leakage, the surfaces can be polished with crocus cloth. DO NOT remove any of the valve material. Replace the valve body and spool if scoring or coating is excessive. If the action of the valve was erratic

or sticky before removal, it may be unbalanced due to wear on the spools or body and should be replaced.

Volume Control Valve Repair

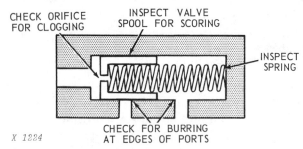

X 1224

Fig. 39 — Inspecting Volume Control Valve

1. On valve spools with orifices, inspect for clogging by dirt or other foreign matter (Fig. 39). Clean with compressed air or a small wire.

2. Rewash all parts thoroughly to remove all emery or metal particles. Any such abrasive could quickly damage the entire hydraulic system.

3. Check the valve spool for freedom of movement in the bore. When lightly oiled, the valve should slide into the bore from its own weight.

Pressure Control Valve Repair

Check for weak relief valve spring with spring tester if system checks have indicated low pressure. This may be remedied by replacing the spring or by adding shims to increase the compression of the spring in some cases. Never add so many shims that the spring is compressed solid.

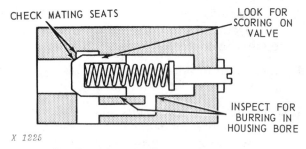

X 1225

Fig. 40 — Inspecting Pressure Control Valve

VALVE SEATS AND POPPETS

Check valve seats for evidence of leak-by and scoring. Replace valve if flat spots appear on seat or on poppets.

Metal valve seats and poppets may be surface polished with crocus cloth if scoring is not deep. **Do not** remove any valve material.

Some seats and valve poppets are made of nylon. This material is long-wearing and sufficiently elastic to conform perfectly to mating surfaces, giving a tight seal.

The nylon seats on poppet valves will take wear, with no damage to the mating metal point. When repairing these valves, always replace nylon parts with new nylon service parts.

CHECKING NON-ADJUSTABLE CARTRIDGE-TYPE RELIEF VALVES

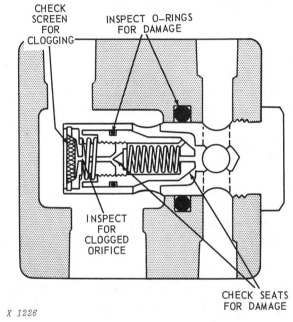

X 1226

Fig. 41 — Inspecting Cartridge-Type Relief Valve

If the relief valve screen or orifice becomes plugged, oil cannot enter the relief valve body to equalize the pressure in the area between the orifice plate and the pilot assembly (Fig. 41).

This plugging then causes the valve to open at lower pressures than it should. The result is sluggish operation of hydraulic units.

Be sure that the relief valve screen and orifice are kept clean at all times.

Also check the O-rings for damage which might cause leakage.

Each relief valve cartridge is stamped with the part number, the pressure limit, and the date of manufacture (Fig. 42). Use this code when testing the cartridges.

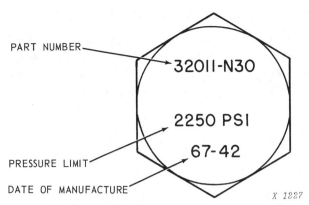

Fig. 42 — How To Read A Cartridge-Type Relief Valve Code

Test the valve cartridges for pressure setting by installing in the system and operating until the valve opening pressure is reached. Read the pressure on a gauge installed in the valve circuit.

ASSEMBLY OF VALVES

1. When assembling valves, be sure that they are kept absolutely clean. Wash parts in kerosene, blow dry with air, then dip in hydraulic oil with rust inhibitor to prevent rusting. This will also aid in assembly and provide initial lubrication. Petroleum jelly can also be used to hold sealing rings in place during assembly.

2. Double check at this time to be sure that valve mating surfaces are free of burrs and paint.

3. Replace all seals and gaskets when repairing a valve assembly. Soak new seals and gaskets in clean hydraulic oil prior to assembly. This will prevent damage and help seal the valve parts.

4. Be sure to insert valve spools in their matched bores. Valve sections must also be assembled in their correct order.

5. When mounting valves, be sure there is no distortion. This may be caused by uneven tension on the mounting bolts and oil line flanges, uneven mounting surfaces, improper location of the valve, or insufficient allowance for line expansion when oil temperature rises. Any of these may result in valve spool binding.

6. After tightening bolts, check the action of the valve spools. If there is any sticking or binding, adjust tension of mounting bolts.

TROUBLE SHOOTING OF VALVES

Listed below are trouble shooting steps which may be used to diagnose most hydraulic valve difficulties. When working on a specific machine, refer to the machine technical manual for more detailed diagnostic information.

PRESSURE CONTROL VALVES

Relief Valves

LOW OR ERRATIC PRESSURE

1. Incorrect adjustment.
2. Dirt, chip or burr holding valve partially open.
3. Worn or damaged poppet or seat.
4. Sticking valve piston in main body.
5. Weak spring.
6. Spring ends damaged.
7. Valve cocking in body or on seat.
8. Orifice or balance hold blocked.

NO PRESSURE

1. Orifice or balance hole plugged.
2. Poppet not seating.
3. Loose fitting of valve.
4. Valve binding in body or cover.
5. Spring broken.
6. Dirt, chip or burr holding valve partially open.
7. Worn or damaged poppet or seat.
8. Valve cocked in body or on seat.

EXCESSIVE NOISE OR CHATTER

1. Too-high oil viscosity.
2. Faulty or worn poppet or seat.
3. Excessive return line pressure.
4. Pressure setting too close to that of another valve in circuit.
5. Improper spring used behind valve.

UNABLE TO ADJUST PROPERLY WITHOUT GETTING EXCESSIVE SYSTEM PRESSURE

1. Spring broken.
2. Spring fatigued.
3. Improper spring.
4. Drain line restricted.

OVERHEATING OF SYSTEM

1. Continuous operation at relief setting.
2. Oil viscosity too high.
3. Leaking at valve seat.

Pressure Reducing Valves

ERRATIC PRESSURE

1. Dirt in oil.
2. Worn poppet or seat.
3. Restricted orifice or balance hole.
4. Valve spool binding in body.
5. Drain line not open freely to reservoir.
6. Spring ends not square.
7. Improper spring.
8. Fatigued spring.
9. Valve needs adjustment.
10. Worn spool bore.

Pressure Sequence Valves

VALVE NOT FUNCTIONING PROPERLY

1. Improper installation.
2. Improper adjustment.
3. Broken spring.
4. Foreign matter on plunger seat or in orifices.
5. Leaky or blown gasket.
6. Drain line plugged.
7. Valve covers not tightened properly or installed wrong.
8. Valve plunger worn or scored.
9. Seat of valve stem worn or scored.
10. Orifices too large, causing jerky operation.
11. Binding from coating of moving parts with oil impurities (due to overheating or improper oil).

PREMATURE MOVEMENT TO SECONDARY OPERATION

1. Valve setting too low.
2. Excessive load on primary cylinder.
3. High inertia load on primary cylinder.

NO MOVEMENT OR SLOWNESS OF SECONDARY OPERATION

1. Valve setting too high.
2. Relief valve setting too close to that of sequence valve.
3. Valve spool binding in body.

Unloading Valves

VALVE FAILS TO COMPLETELY UNLOAD PUMP

1. Valve setting too high.
2. Pump failing to build up to unloading valve pressure.
3. Valve spool binding in body.

DIRECTIONAL CONTROL VALVES

Spool Valves
Rotary Valves
Check Valves

FAULTY OR INCOMPLETE SHIFTING

1. Worn or binding control linkage.
2. Insufficient pilot pressure.
3. Burned-out or faulty solenoid.
4. Defective centering spring.
5. Improper spool adjustment.

ACTUATING CYLINDER CREEPS OR DRIFTS

1. Valve spool not centering properly.
2. Valve spool not shifted completely.
3. Valve spool body worn.
4. Leakage past piston in cylinder.
5. Valve seats leaking.

CYLINDER LOAD DROPS WITH SPOOL IN CENTERED POSITION

1. Lines from valve housing loose.
2. O-rings on lockout springs or plugs leaking.
3. Broken lockout spring.
4. Circuit relief valves leaking.

CYLINDER LOAD DROPS SLIGHTLY WHEN RAISED

1. Defective check valve spring or seat.
2. Spool valve position improperly adjusted.

OIL HEATS (CLOSED-CENTER SYSTEMS)

1. Valve seat leakage (pressure or return circuits).
2. Improper adjustment of valves.

VOLUME CONTROL VALVES

Flow Control and Flow Divider Valves

VARIATIONS IN FLOW

1. Valve spool binding in body.
2. Leakage in cylinder or motor.
3. Oil viscosity too high.
4. Insufficient pressure drop across valve.
5. Dirt in oil.

ERRATIC PRESSURE

1. Worn valve poppet or seat.
2. Dirt in oil.

IMPROPER FLOW

1. Valve not adjusted properly.
2. Restricted valve piston travel.
3. Restricted passages or orifice.
4. Cocked valve piston.
5. Circuit relief valve leaking.
6. Oil too hot.

OIL HEATS

1. Improper pump speed.
2. Holding hydraulic functions in relief.
3. Improper connections.

TEST YOURSELF

QUESTIONS

I. Introduction

1. Name the three basic valve types.

II. Pressure Control Valves

1. Give a couple of reasons why pressure control valves are used in hydraulic systems.

2. Name the two general types of relief valves.

3. Compare cracking pressure and full flow pressure.

4. At which of these pressures are relief valves rated?

5. Why are direct-acting relief valves considered to be safe should they fail in operation?

6. Compare a pressure *relief* valve with a pressure *reducing* valve.

III. Directional Control Valves

1. Name the five types of valving elements in directional control valves.

2. What is least complicated and most often used of all directional control valves?

3. What are the two basic types of sliding spool valves?

4. What is the difference in oil flow through the two spool valves in neutral?

IV. Volume Control Valves

1. What two methods are used to control volume flow?

2. What are "noncompensated" and "compensated" flow control valves?

3. What is the simplest type of flow control valve?

4. How do you control flow through a pressure-compensated flow control valve?

V. Microprocessor-based Electronic Control of Hydraulic Valves.

1. What are the four advantages of microprocessor-based electronic control of hydraulic valves?

HYDRAULIC CYLINDERS / CHAPTER 4

INTRODUCTION

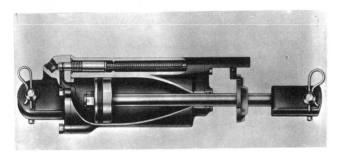

Fig. 1—Piston-Type Cylinder

The cylinder does the work of the hydraulic system. It converts the fluid power from the pump back to mechanical power. Cylinders are the "arms" of the hydraulic circuit.

Chapter 1 explains the uses of hydraulics and shows how cylinders can be used to actuate both mounted equipment and drawn implements (remote uses). In either case, the basic design of the cylinder is the same; only the extra features are different.

TYPES OF CYLINDERS

Two major types of cylinders are covered in this chapter:

- **Piston-Type Cylinders—give straight movement.**

- **Vane-Type Cylinders—give rotary movement.**

(Another type of rotary actuator is the hydraulic motor, covered in Chapter 5.)

PISTON-TYPE CYLINDERS

Two major kinds of piston-type cylinders are used:

• SINGLE-ACTING CYLINDERS—give force only one way (Fig. 2). Pressure oil is admitted to only one end of the cylinder, raising the load. An outside force such as gravity or a spring must return the cylinder to its starting point.

• DOUBLE-ACTING CYLINDERS—give force in both directions (Fig. 2). Pressure oil is admitted first at one end of the cylinder, then at the other, giving two-way power.

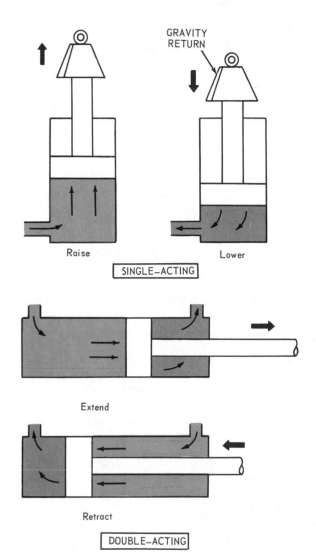

X 1124

Fig. 2—Single- and Double-Acting Cylinders Compared

In both types of cylinders, a movable piston (or rod) slides in a cylinder housing or barrel in response to pressure oil admitted to the cylinder. The piston may use various packings or seals to prevent leakage.

The force (F) exerted by a piston can be determined by multiplying the piston area (A) by the pressure (P) applied.

Force = Pressure x Area

Fig. 3 — Force Triangle

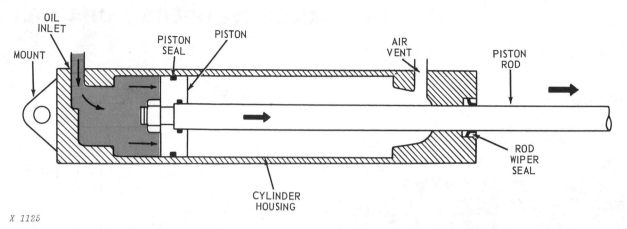

Fig. 4 — A Typical Single-Acting Cylinder

SINGLE-ACTING CYLINDERS

In a single-acting cylinder, pressure oil is applied to only one side of the piston (Fig. 4). The piston and rod are forced out of the housing as shown, moving the load. When the oil pressure is released, the weight of the load (or a spring device) forces the rod back into the housing. The cylinder mount holds the cylinder in place while it works.

The other side of the cylinder is dry. A small air vent is required to release air when the piston rod extends, and to let in air when the rod retracts. This allows the cylinder to work smoothly and prevents a vacuum. To keep out dirt, a porous breather is often used in the air vent.

A seal on the piston prevents leakage of oil into the dry side of the cylinder. A wiper seal in the rod end of the cylinder cleans the rod as it moves in and out of the housing.

In some single-acting cylinders, the piston rod has no piston on the inner end. Instead, the end of the rod serves as the piston. This is called a **ram**-type cylinder (Fig. 5). The rod is slightly smaller than the inside of the cylinder. (A small shoulder or ring on the end of the rod keeps the rod from being pushed out of the cylinder.)

The ram-type construction has several advantages over the piston-type: 1) The rod is bigger and resists bending due to side loads. 2) The packing is on the outside and is easier to reach. 3) Scoring inside the cylinder bore will not damage packings. 4) No air vent is needed since oil fills the whole inner chamber of the cylinder housing.

Single-acting cylinders are favored on some mobile equipment where a simple hydraulic lift is needed and the weight of the working unit will lower itself.

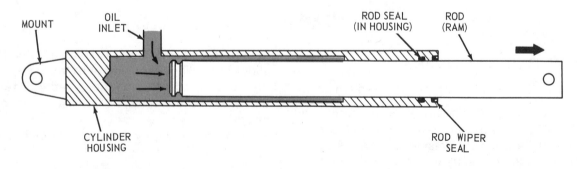

Fig. 5 — Ram-Type Cylinder

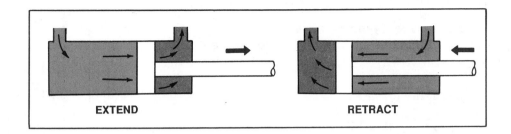

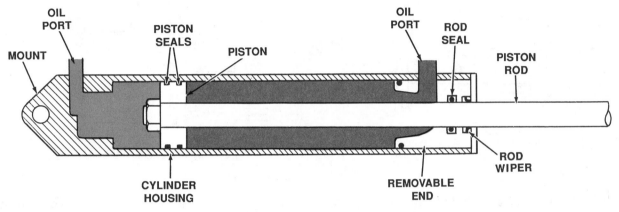

Fig. 6 — A Typical Double-Acting Cylinder

DOUBLE-ACTING CYLINDERS

Double-acting cylinders provide force in both directions. Pressure oil enters at one end of the cylinder to extend it, at the other to retract it (Fig. 6). Oil from the opposite end of the cylinder returns to reservoir each time.

With the double-acting cylinder, both the piston head and the piston rod must be sealed to prevent oil leakage.

Two types of double-acting cylinders are shown in Fig. 7.

In the UNBALANCED or differential type, total force on the rod side of the piston is less than that on the blank side. This is because the rod fills in an area not exposed to pressure. This cylinder is usually designed for a slower, more powerful stroke when it extends, and for a faster, less powerful stroke when it retracts.

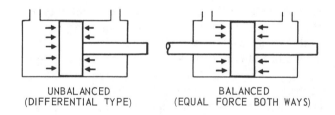

X 1128

Fig. 7 — Two Types of Double-Acting Cylinders

In the BALANCED cylinder, the piston rod extends through the piston head on both sides. This gives equal working area on both sides of the piston and balances the working force of the cylinder whether it is extending or retracting.

(Of course, the balance or unbalance of these cylinders depends also upon the loads. If the cylinder is not attempting to move equal loads in each direction, the balances will vary.)

EXTRA FEATURES OF PISTON-TYPE CYLINDERS

Many piston-type cylinders have extra features which add functions or adapt them to different uses.

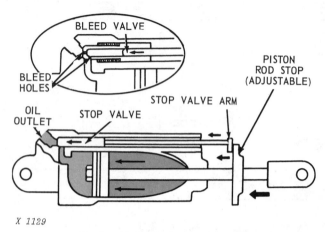

X 1129

Fig. 8 — Hydraulic Stop in Cylinder

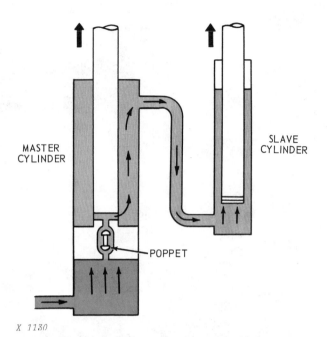

X 1130

Fig. 9 — Use of Master and Slave Cylinders

STROKE CONTROL DEVICES

A hydraulic stop is sometimes used to stop the piston at any point of its travel by shutting off the flow of outgoing oil (Fig. 8).

As the piston retracts, the rod stop contacts the arm as shown, moving the stop valve against its seat and restricting the oil outlet. The trapped oil resists the piston and causes oil pressure to rise on the opposite end. This pressure rise affects the control valve and returns the system to neutral. The piston rod stop is adjustable for any stroke.

Mechanical stops are also used to stop some cylinders at a preset point on their strokes.

An "override" feature is built into the cylinder shown in Fig. 8. After the stop valve closes, two small bleed holes in the valve (see inset) allow a limited flow of oil out of the cylinder as the piston is retracted further (by holding the control lever in retract position). This action moves a bleed valve inside the stop valve (see inset) until it finally seats in the end of the stop valve, stopping all oil flow from the cylinder.

A spring mechanism allows incoming oil to reopen the valves for the next stroke of the cylinder.

The rate of operation of some cylinders is adjustable. Usually this is done at the control valve for the cylinder by means of a volume control device (see Chapter 3 for details).

SLAVE CYLINDERS

A slave cylinder can be used, supplied by a master cylinder (Fig. 9). Oil flows to the master cylinder, actuating it. Some of this pressure oil flows through a poppet or orifice in the piston and on into the opposite end of the cylinder. This oil then flows to the slave cylinder, actuating it in turn. The poppet or orifice can be designed so that both cylinders operate in unison or by delayed action. When the master cylinder is retracted, a matching poppet in the master cylinder opens, allowing oil to flow back through the master piston and return to reservoir.

In a variation on the slave cylinder setup, three cylinders are connected in series. Oil is fed to the rod end of largest cylinder, actuating it. This cylinder then pumps oil to the second largest cylinder, which in turn serves the smallest cylinder. Cylinder displacements can be matched so that all three cylinders operate at once. A common use is in controlling three sections of drawn implements. However, the largest cylinder must be capable of lifting the entire implement.

CUSHIONS

A cushion is built into some cylinders to slow them down at the end of their strokes. This cushion is used as a "hydraulic brake" to protect against impact damage. In Fig. 10, the cylinder works normally during its main stroke (top), but slows down as the piston seals off the oil outlet (bottom). Now the outlet oil must go through the small orifice, slowing the piston.

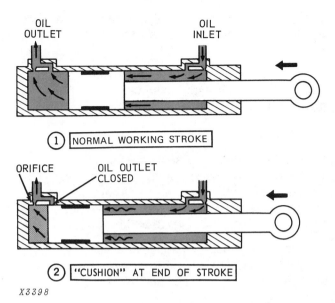

Fig. 10 — Use of "Cushion" in Cylinder

STEPPED PISTONS

A stepped piston allows a cylinder to provide a rapid starting stroke with low force and a slower, more powerful working stroke. This is done by admitting oil first against the smaller part of the piston, which moves rapidly until the work is contacted (Fig. 11). Then the entire piston surface takes over for the power stroke.

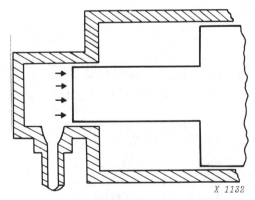

Fig. 11 — Use of Stepped Piston in Cylinder

REGENERATING CYLINDERS

In a regenerating cylinder, oil discharged from the rod end is routed back to the head end to help speed up the stroke. Return oil from the rod end of the cylinder is routed to the piston end as shown in Fig. 12 and adds its volume to the normal flow through the control valve to the cylinder. During this cycle, pressure is equal on both the head and rod end of the cylinder. However, the cylinder still extends because the areas of the two ends are not equal. In effect, pressure is applied to a larger

area at the left end, moving the piston to the right as shown. The piston will move very fast, but will have little force. (Pressure x area of rod cross-section).

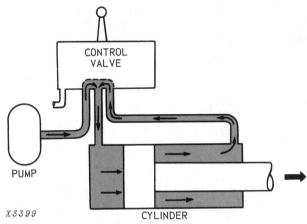

Fig. 12 — Regenerating Cylinder (Double-Acting)

PISTON-TYPE CYLINDERS WHICH PRODUCE ROTARY MOTIONS

Two piston-type cylinders mounted opposite each other and operated as a push-pull device can be used to produce limited rotary motion. (Vane-type cylinders also produce rotary motion—see later in this chapter.) In some cases, a rack and pinion assembly is adapted for this, using a single-acting cylinder at each end of the rack. As pressure is applied at one end, the rack slides in its housing, turning the pinion. "Cushions" (see above) are often used to slow down the motion at the end of the piston stroke.

TELESCOPING PISTONS

In this application, the cylinder rod has an inner and outer section. The rod acts as one piece until the outer section reaches its stop. Then the inner section continues to the end of its stroke. They may be designed so that the inner section moves first, followed by the outer section. In this case, a locking device is used between the two sections. The speed of each section depends upon the area of the pushing surface of each.

PROTECTIVE CHECK VALVES

Some cylinders use a check valve at the oil inlet to protect the cylinder against loss of fluid from a line failure or other leak. If the oil supply fails, the check valve closes and traps the oil in the cylinder. This is very important where a heavy load is riding on the cylinder. An example is the leveling cylinders on a hillside combine. (See Chapter 3 for details on "Check Valves.")

PROTECTIVE THERMAL RELIEF VALVES

Heat can cause oil to expand and raise the pressure in a cylinder. Sometimes the heat of the sun can rupture a cylinder at rest. To prevent this, some cylinders have a thermal relief valve which is set far higher than system pressure and acts as a safety valve for this high pressure oil. (See Chapter 3 for details.)

VANE-TYPE CYLINDERS

A vane-type cylinder provides rotary motion.

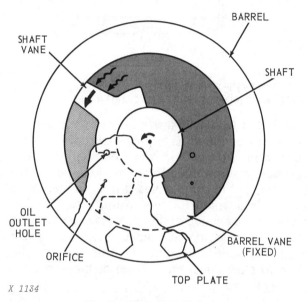

X 1134

Fig. 13 — Vane-Type Cylinder

Fig. 13 shows a vane-type cylinder in operation. In a round barrel, the shaft and vane rotate as pressure oil enters. Oil is discharged through the outlet hole in the other side of the cylinder.

A "cushion" or "hydraulic brake" can be built into the vane-type cylinder as shown. As the shaft vane travels, it shuts off the oil outlet hole in the top plate. This leaves only a small orifice to discharge the oil, slowing the rotating vane as it comes to the end of its stroke.

The vane-type cylinder is used to swing rotary equipment such as a backhoe. It allows the operator to swing the boom and bucket rapidly from trench to pile and back again. The optional "hydraulic brake" prevents jerky stops and impact damage.

Most vane-type cylinders are double-acting, as shown. The fixed barrel vane separates the two chambers of the cylinder. Pressure oil is sent first to one chamber to swing left, then to the other to swing right.

Limited rotary movement can also be devised by using two piston-type cylinders in a push-pull arrangement (see earlier in this chapter).

SEALS

Many different types of seals are used in cylinders. The seal used depends upon the oil pressure and temperature, whether the parts are moving or static, the speed of the moving parts, and the amount of shock loading. See Fig. 14.

Chapter 9 of this manual covers seals in detail. Refer to this chapter for information on types and uses of seals, and correct maintenance.

IDENTIFYING SIZE OF CYLINDERS

Cylinders on mounted equipment are sized for the job by the manufacturer.

However, with remote cylinders, there is a chance that the wrong cylinder may be used on a certain job.

Labels are used on most of these cylinders to identify the size. A typical label can be read as follows:

Label		Cylinder Diameter
25	64	2½"
30	76	3"
35	89	3½"

The first two digits on the label give the diameter of the cylinder in tenths of inches. Merely place a decimal point between the two digits and you have the size of the cylinder in inches. For example "25" = 2.5 inches. (The last two digits on the label give the size in millimeters.)

If you are not sure what size cylinder to use for a job, check the operator's manual for the machine or check the part's number in the parts catalog.

TESTING AND DIAGNOSING CYLINDER PROBLEMS

Cylinders can be tested on the machine for leaks and other failures. Refer to Chapter 12, "Testing and Diagnosis," for details. This chapter also has charts which list the causes and remedies for each possible failure.

The following section on "Maintenance of Cylinders" covers a few of the common problems.

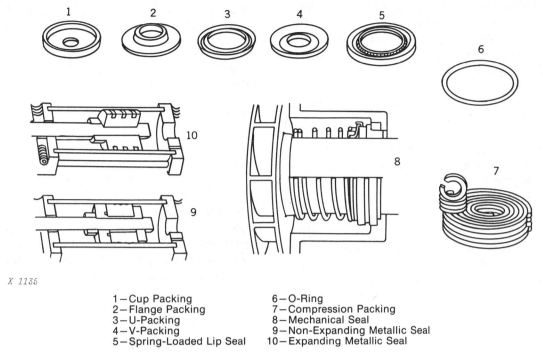

X 1135

1—Cup Packing
2—Flange Packing
3—U-Packing
4—V-Packing
5—Spring-Loaded Lip Seal

6—O-Ring
7—Compression Packing
8—Mechanical Seal
9—Non-Expanding Metallic Seal
10—Expanding Metallic Seal

Fig. 14 — Types of Seals for Hydraulic Cylinders

MAINTENANCE OF CYLINDERS

Hydraulic cylinders are compact and relatively simple. The key points to watch are the seals and the pivots. Here are a few service tips:

1. EXTERNAL LEAKAGE—If the cylinder end caps are leaking, tighten the caps. If this fails to stop the leak, replace the gasket. If the cylinder leaks around the piston rod, replace the packing. Be sure the seal lip faces toward the pressure oil. If the seal continues to leak, check items 5 through 9 below.

2. INTERNAL LEAKAGE—Leakage past the piston seals inside the cylinder can cause sluggish movement or settling under load. Piston leakage can be caused by worn piston seals or rings, or scored cylinder walls. The latter may be caused by dirt and grit in the oil.

IMPORTANT: When repairing a cylinder, be sure to replace all seals and packings before reassembly.

3. CREEPING OF CYLINDER—If the cylinder creeps when stopped in midstroke, check for internal leakage (item 2). Another cause could be a worn control valve (see Chapter 3 for details).

4. SLUGGISH OPERATION—Air in the cylinder is the most common cause of sluggish action. (To bleed air, see end of this chapter). Internal leakage in the cylinder is another cause (item 2). If action is sluggish when starting up the system, but speeds up when the system is warm, check for oil of too-high viscosity (see machine operator's manual). If the cylinder is still sluggish after these checks, the whole circuit should be tested for worn components (see Chapter 12, "Testing and Diagnosis," for details).

5. LOOSE MOUNTING—Pivot points and mounts may be loose. The bolts or pins may need to be tightened or they may be worn out. Too much "slop" or "float" in the cylinder mountings damages the piston rod seals. Check all cylinders for loose mountings periodically.

6. MISALIGNMENT—Piston rods must work in line at all times. If they are "side loaded," the piston rods will be galled and the packings will be damaged, causing leaks. Eventually the piston rods may be bent or the welds broken.

7. LACK OF LUBRICATION—Lack of piston rod lubrication may cause the rod packing to seize, resulting in an erratic stroke, especially on single-acting cylinders.

8. ABRASIVES ON PISTON ROD—When piston rods extend, they can pick up dirt and other material. Then when the rod retracts, it carries the grit into the cylinder, damaging the rod seal. This is why rod wipers are often used at the rod end of the cylinder to clean the rod as it retracts. Rubber

boots are also used over the end of the cylinder in some cases. Another problem is rusting of piston rods. When storing cylinders, always retract the piston rods to protect them. If they cannot be retracted, coat them with grease.

9. BURRS ON PISTON ROD—Exposed piston rods can be damaged by impact with hard objects. If the smooth surface of the rod is marred, the rod seal may be damaged. Burrs on the rod should be cleaned up immediately using crocus cloth. Some rods are chrome plated to resist wear. Replace seals after rod surface is restored.

10. CHECKING AIR VENTS—Single-acting cylinders (except ram-types) must have an air vent in the dry side of the cylinder. To prevent dirt entry, various filter devices are used. Most are self-cleaning, but they should be inspected periodically to insure proper operation.

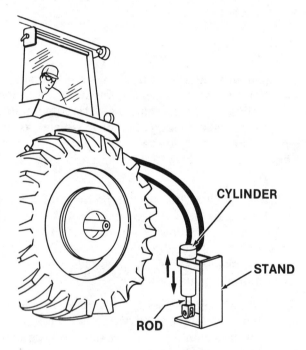

Fig. 15 — Bleeding Air From Remote Cylinder

BLEEDING AIR FROM REMOTE CYLINDERS

Any time a remote cylinder is plugged into the hydraulic circuit, all trapped air must be bled. This will prevent sluggish action of the cylinder.

First attach the cylinder to the circuit. Place the cylinder on the ground (or on a hanger) with the piston rod end down as shown in Fig. 15. (Or on mounted cylinders, place the head end of the cylinder in its working mount, allowing the rod end freedom to move in and out.) Start the machine and move the hydraulic control lever back and forth seven or eight times to extend and retract the cylinder. This will bleed the air. (On double-acting cylinders, you may have to turn the cylinder end-for-end and repeat the cycling of the control lever.) Do not stand near the cylinder when it is being bled.

TEST YOURSELF

QUESTIONS

1. (True or false?) "Cylinders convert mechanical power to fluid power."

2. (Fill in the blanks.) Piston-type cylinders give _____ movement while vane-type cylinders give _____ movement.

3. (Fill in the blanks.) Cylinders which are _____-acting give force in both directions. Cylinders which are _____-acting give force in only one direction.

4. What fills the chambers on each side of the piston in a single-acting cylinder?

5. In a double-acting cylinder, how does a piston rod on one side only affect the working stroke?

6. How does a "hydraulic brake" or "cushion" work in a vane-type cylinder?

HYDRAULIC MOTORS / CHAPTER 5

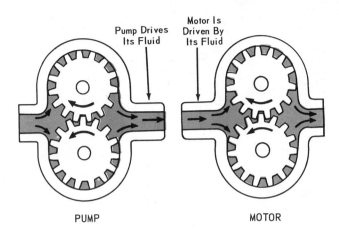

PUMP   MOTOR

X 1296 Fig. 1—Hydraulic Pump and Motor Compared

A hydraulic motor works in reverse when compared to a pump (Fig. 1).

The pump *drives* its fluid, while the motor is *driven by* its fluid. Thus:

• **Pump—draws in fluid and pushes it out, converting mechanical force into fluid force.**

• **Motor—fluid is forced in and exhausted out, converting fluid force into mechanical force.**

This combination is also referred to as a hydrostatic drive system.

In use, the pump and motor are often hydraulically coupled to provide a power drive:

1. The pump is driven mechanically, drawing in fluid and pumping it to the motor.

2. The motor is driven by the fluid from the pump and so drives its load by a mechanical link.

The motor is really an actuator, like the cylinder (Chapter 4). However, the motor is a rotary actuator that rotates in a full circle. (The vane-type cylinder is a *limited* rotary actuator.)

COMPARING PUMP AND MOTOR DESIGN

The motor is designed much like the pump. Both use the same basic types—gear, vane, and piston. Often their parts can be substituted one for the other.

Both pump and motor use an internal sealing of parts to back up their flow of fluid—*positive* displacement. Without this seal, a motor's elements would not move under force of the incoming fluid.

Sometimes pumps are modified and used as motors. However, a pump should never be used as a motor or converted to a motor without considering all the factors of the application. For example, the wear on shaft bearings often increases in motor usage.

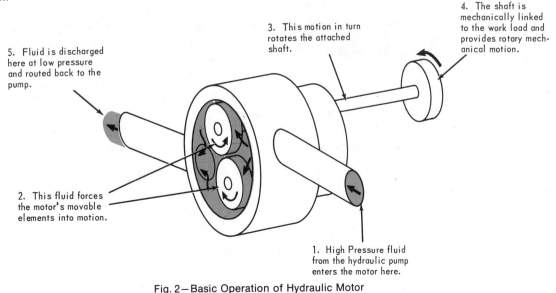

5. Fluid is discharged here at low pressure and routed back to the pump.

3. This motion in turn rotates the attached shaft.

4. The shaft is mechanically linked to the work load and provides rotary mechanical motion.

2. This fluid forces the motor's movable elements into motion.

1. High Pressure fluid from the hydraulic pump enters the motor here.

X 1297

Fig. 2—Basic Operation of Hydraulic Motor

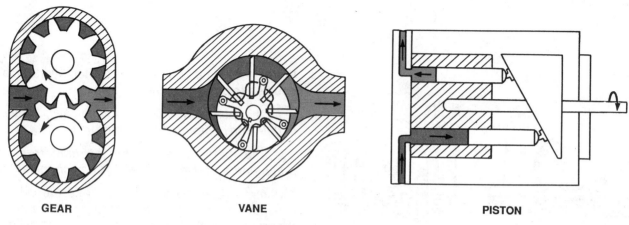

GEAR   VANE   PISTON

Fig. 3—Three Types of Hydraulic Motors

DISPLACEMENT AND TORQUE OF MOTORS

The work output of a motor is called *torque*. This is a measure of the rotary force on the motor drive shaft.

Torque is only a measure of force X distance (as in "foot-pounds") not of the *speed* of this force.

The ratio between the speed and torque output of a motor depends upon its *displacement*—the volume of fluid it displaces with each cycle.

Motors, like pumps, are designed for two types of displacement:

• FIXED DISPLACEMENT motors normally have variable speeds which are regulated by varying the amount of input flow. Normally these motors have a fixed torque, or rotary work output.

• VARIABLE DISPLACEMENT motors have both variable speeds and variable torques. The input flow and pressure remain constant, while the speed and torque can be varied by mechanisms which change the displacement.

The applications and efficiencies of these motors will be discussed later in this chapter.

TYPES OF HYDRAULIC MOTORS

Motors are designed in three basic types (Fig. 3):

• **Gear Motors**

• **Vane Motors**

• **Piston Motors**

These basic types are the same as for the pump (Chapter 2).

All three designs work on the rotary principle: a rotating unit inside the motor is moved by the incoming fluid.

Let's discuss the operation of each type of motor.

GEAR MOTORS

Gear motors are widely used because they are simple and economical. Often they are used to drive small equipment in remote applications.

Usually small in size, gear motors are versatile and can be transferred from one use to another by using a universal mounting bracket and flexible hoses.

Gear motors can rotate in either direction but are normally not capable of variable displacement.

Two basic designs are used:

• **External Gear Motors**

• **Internal Gear Motors**

Let's discuss the operation of each.

EXTERNAL GEAR MOTORS

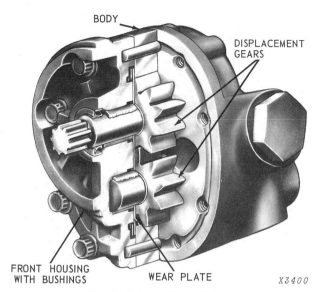

Fig. 4—External Gear Motor

The external gear motor is a duplicate of the external gear pump. It has two equal-sized gears in mesh, sealed in a housing (Fig. 4).

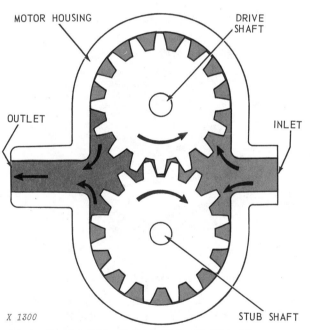

Fig. 5—External Gear Motor in Operation

In operation, pressure oil from the pump forces the motor gears to rotate away from the inlet port. This action rotates the motor shaft connected to the work load.

The force of the pressure oil is expended as it travels between gear teeth and housing to the outlet port. Here it leaves the motor as low pressure fluid and returns to the reservoir or the pump.

Balanced Version of External Gear Motor

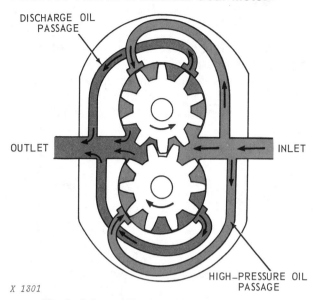

Fig. 6—Balanced Version of External Gear Motor

Some external gear motors are "balanced" for equal pressure on all sides of the rotating parts (Fig. 6). This is done to reduce bearing failures. The basic motor is the same but passages have been added in the housing to connect inlet and outlet oil pressure to the opposite sides of the motor. Now pressure is exerted equally on both sides of the gears and shafts, cancelling out unequal pressures and balancing the motor.

INTERNAL GEAR MOTOR

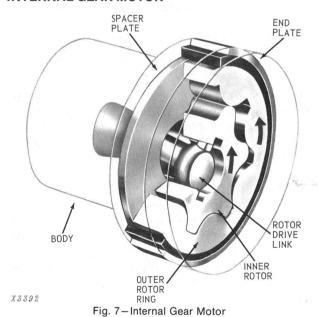

Fig. 7—Internal Gear Motor

One popular internal gear motor is much like the rotor pump (Chapter 2). This motor is shown in Fig. 7.

Not actually gears, the moving parts are called the rotor and the rotor ring. The rotor is driven inside the rotor ring.

The rotor is mounted eccentric to the rotor ring. The ring has one more lobe than the rotor so that only one lobe is in full engagement with the outer ring at any one time. This allows the rotor's lobes to slide over the outer lobes, making a seal.

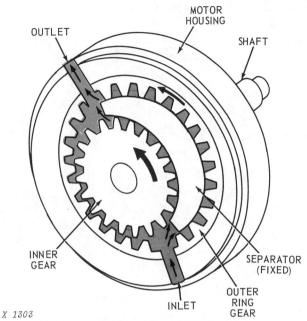

Fig. 9—Internal Gear Motor With Separator

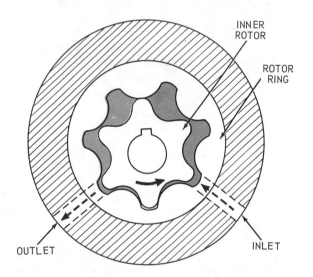

Fig. 8—Internal Gear Motor in Operation

In operation, pressure fluid enters the motor, striking the rotor and rotor ring lobes, forcing both to rotate (Fig. 8). As they rotate, a seal is formed, then broken, as each inner lobe engages a cavity in the outer ring. The fluid is discharged under low pressure at the outlet port.

Fig. 9 shows another type of internal gear motor. It uses a crescent-shaped separator between the inner and outer gears—just as the internal gear pump in Chapter 2. Operation is basically the same, but with inlet and outlet pressures reversed.

VANE MOTORS

Vane motors, like their "brother" pumps, are available in two types—balanced and unbalanced. Most of the vane motors on today's machines are the balanced type, since most of these applications do not require variable displacement. Balanced motors have a longer service life (less bearing wear) and so are more economical.

The balanced vane motor (Fig. 11) operates much the same as the vane pump (Chapter 2). The slotted rotor turns, driven in this case by the force of incoming oil against the vanes.

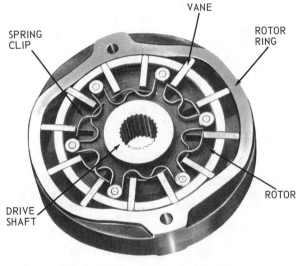

Fig. 10—Vane Motor (Balanced Type)

The only extra feature of the motor is the devices used to hold the vanes in contact with the outer ring. These devices may be spring clips as shown in Fig. 10 or small springs beneath each vane pushing it out.

These devices are needed for internal sealing in a motor but not in a pump. In the pump, centrifugal force throws the vanes out against the outer ring. But in the motor, incoming oil is under high pressure and would bypass the vanes before rotation began unless the vanes were held out solid against the ring.

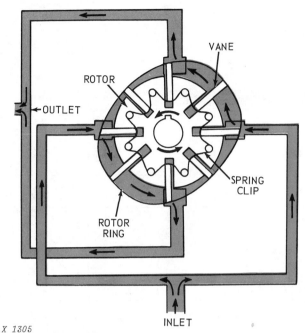

X 1305

Fig. 11—Balanced Vane Motor in Operation

SUMMARY: BALANCED VANE MOTORS

Balanced vane motors are only capable of fixed displacement. Yet they usually provide more power and efficiency than gear motors. Their direction of rotation can easily be changed by reversing the flow of fluid.

NOTE: Most vane pumps can be converted to motors, but unless a vane pump has clips or springs to hold the vanes in contact with the rotor ring, it will not work as a motor.

PISTON MOTORS

Piston motors are often favored in systems which have high speeds or high pressures. While more sophisticated than the other two types, piston motors are more complex, more expensive, and require more careful maintenance.

Like its pump counterpart, the piston motor is available in two types:

• **Axial Piston**

• **Radial Piston**

On mobile systems, axial piston models are often favored. The radial piston model is usually confined to stationary industrial uses where space is not limited and more power is needed.

AXIAL PISTON MOTORS

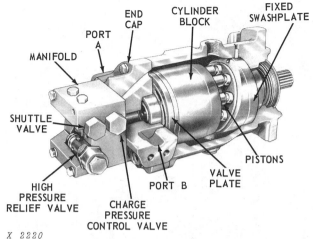

X 2220

Fig. 12—Axial Piston Motor (Fixed Displacement)

Fig. 12 shows a fixed displacement version of an inline axial piston motor.

The motor shown is used as part of the hydraulic drive on a self-propelled machine.

The end cap contains ports A and B which accept pressure oil from the pump to operate the motor. They also discharge low pressure oil back to the pump.

The pistons work in bores in the revolving cylinder block and contact a fixed-angle swashplate as they revolve.

In operation (Fig. 13), high pressure oil through port A enters cylinder bores, forcing its piston against the angled swashplate. Because the swashplate is fixed, the piston slides down its angled face. This sliding action makes it revolve, turning the cylinder block which in turn drives a shaft which propels the load.

As the cylinder block turns, other bores align with port A and their pistons are actuated, keeping up the rotation.

During the second half of the motor's revolution, low pressure oil is discharged at port B as the pistons are forced back by the thicker part of the swashplate.

To reverse the rotation, simply reverse the flow of oil, feeding pressure oil in at port B and returning it through port A.

The valving shown at left in Fig. 13 helps to control and protect the motor. A flow divider valve, two high-pressure relief valves, and a pressure control valve are used.

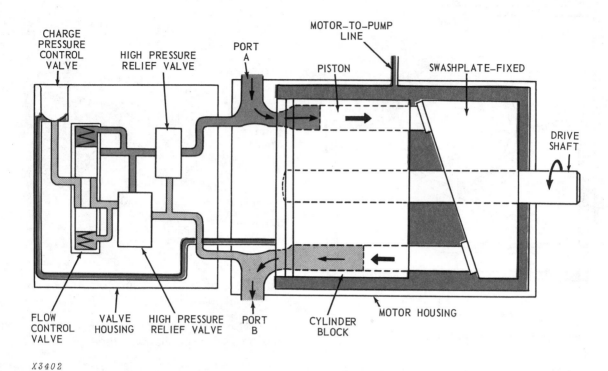

Fig. 13—Axial Piston Motor in Operation (Fixed Displacement)

The basic operation of these valves is given in Chapter 3. For the motor shown, the valves work as follows:

The *flow divider valve* moves back and forth in response to incoming pressure oil and prevents this oil from entering the low pressure side of the circuit. It keeps the incoming oil open to the charge pressure control valve while the motor is "in stroke".

The *high pressure relief valves* monitor the incoming pressure oil at each port—whenever that port is the inlet route for oil. When pressure exceeds the valve setting, it opens and bypasses oil, slowing or stopping the motor. This protects the motor against overloads. When pressures are again lowered to normal, the valve closes and the motor speeds up as inlet oil flows in again.

The *charge pressure control valve* routes excess oil from the system charging pump to the motor housing and back to the main pump. This oil aids in cooling and lubricating the motor and main pump.

Variable Displacement Version of Axial Piston Motor

This motor is shown in Fig. 14. It is used as part of a pump-motor combination to drive a self-propelled machine.

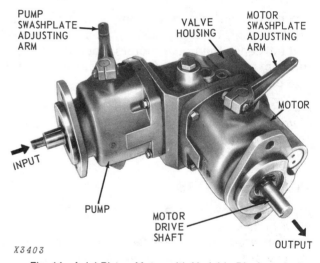

Fig. 14—Axial Piston Motor with Variable Displacement (Pump-Motor Drive Unit Shown)

The pump and motor share common valving and both are connected at a 90° angle as shown. The valving sends fluids from pump to motor. A small charging pump is also built into the valve housing.

The pump and motor are very similar; most parts are identical.

IMPORTANT: *Although many pump and motor parts are identical, do not interchange them after the unit has been in service. Moving parts tend to set up their own peculiar wear patterns and these may not conform to another component. The result might be internal leakage and loss of efficiency.*

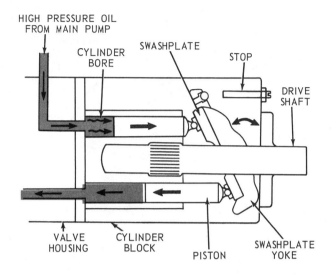

X 1309

Fig. 15—Operation of Axial Piston Motor with Variable Displacement

Operation is shown in Fig. 15. This motor resembles the fixed displacement version described above. The main difference is that the former had a fixed-angle swashplate while the latter has an adjustable swashplate.

Operation is much the same as for the fixed displacement model. High pressure oil forces the pistons to contact and slide down the face of the swashplate, revolving the cylinder block and drive shaft. Fluid at low pressure is expelled as the pistons are forced back by the thicker part of the swashplate.

The displacement of oil in each cycle of the motor is determined by how far the piston must travel to contact the swashplate.

In Fig. 15, the angle of the swashplate can be adjusted by tilting it using an arm and lever. By moving the lever, the displacement of the motor is changed.

A mechanical stop is used in this motor to limit the distance the swashplate angle can be changed.

The greater the angle of the swashplate, the more oil that is displaced and the faster the motor will normally drive its load. (However, if a fixed displacement pump is supplying the motor, the result will be greater torque but slower speed.)

CAM LOBE MOTORS

The cam lobe motor is a variation of piston type motors.

The cam lobe motor comes in two types:

- **Steerable Motor**
- **Fixed-Mount Motor**

The steerable motor is generally used in vehicle applications. The fixed-mount motor is used in vehicle applications for fixed axles as well as other applications.

On the next pages we will look at the fixed-mount motor since the basic elements are identical between the motors.

OPERATION OF CAM LOBE MOTOR

The power stroke is developed when pressurized oil flows through the manifold and carrier oil passages, forcing each piston outward.

As the piston follower is forced against the cam ramp, the carrier is forced to turn as the follower moved down the cam ramp (Fig. 16).

Pressurized oil is routed through the inner cover to the manifold. Each passage in the manifold is timed to one cam ramp. There are 15 pressure oil and 15 return oil passages in the manifold.

The oil manifold is pressed against the piston carrier with a thin film of oil separating the manifold and carrier surfaces. The piston carrier is splined to the axle. The pressurized oil if fed from the pressure passages in the manifold through passages in the carrier and into the piston bores.

Return oil is routed from piston bores, through the carrier passages and into the manifold return passages.

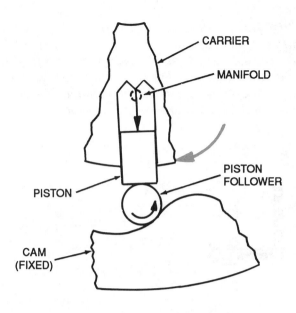

Fig. 16 — Piston, Piston Follower, and Cam

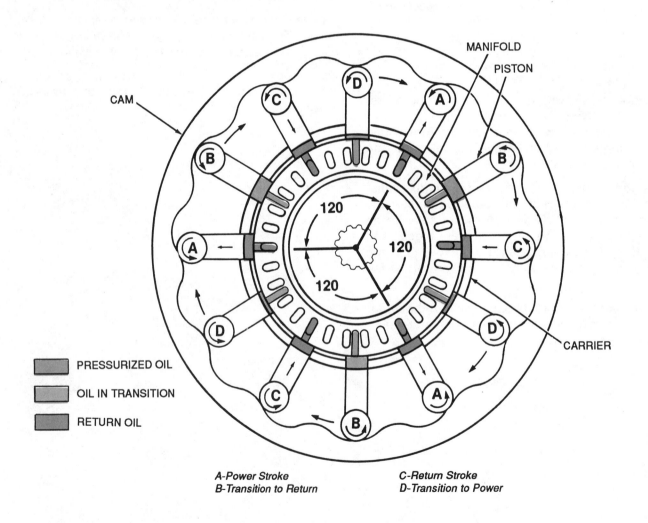

PRESSURIZED OIL

OIL IN TRANSITION

RETURN OIL

A-Power Stroke
B-Transition to Return

C-Return Stroke
D-Transition to Power

Fig. 17—Piston Operation

Piston Operation

The cam lobe motor is designed so there are three pistons (in a 12-piston motor) doing the same thing at equal (120°) distances around the motor (Fig. 17). This gives a force balance on the carrier and axle assembly.

In Fig. 17, with the motor engaged, one group of pistons (A) is in the power stroke. As the piston follower is forced down the cam ramp, the carrier is forced to turn. The group of pistons (C) push the return oil back through the manifold.

Oil is in transition from one point to another. As the carrier turns, pistons (D), at the top of the lobes, begin their power stroke and the pistons (B), at the bottom, begin their return stroke.

As one group ends its power stroke, another group has begun. This overlapping of strokes, both power and return, serve to smooth the power (torque) applied to the axle.

OPERATION OF FIXED-MOUNT MOTOR

The internal grooves (pressure ports) within the inner cover are not connected together as on steerable motors. An external control valve routes high-pressure oil through a porting block as shown on Fig. 18.

The 3-speed porting block maintains the separation between the pressure ports in the inner cover, depending on selected speed. The single-speed porting block combines both pressure ports in the inner cover to provide only the single (low) speed.

Low Speed

Low speed as shown in Fig. 18 occurs when oil flow is routed to both the inner and outer grooves. Oil flow is directed to the greatest number of pistons at any given moment, thus producing the maximum amount of touque.

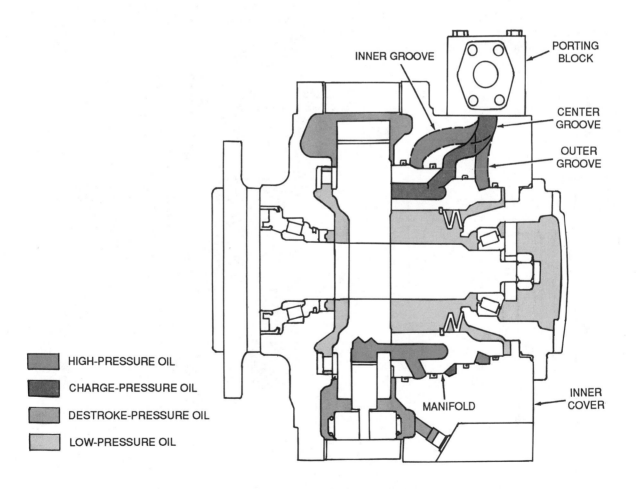

Fig. 18 — Fixed Mount Motor — Low Speed

Middle Speed

Middle speed occurs when oil flow is routed only to the outer groove. High-pressure oil is routed to only 10 of the 15 ports in the manifold. The axle will turn one and a half revolutions using the same volume of oil required for one revolution in low speed. The pistons not involved are recirculating return oil from the return circuit via an external valve.

High Speed

High speed occurs when pressure oil is routed only to the inner groove. This groove feeds only five manifold passages which lead to the carrier pistons. High speed develops the lowest torque because there are 1/3 as many power strokes as compared to low speed. The axle will turn three revolutions using the same volume of oil required in low speed.

Reverse

Reversing the oil flow for 3-speed motors does NOT reverse all three speeds. There is no high-speed reverse. Low-speed reverse is accomplished when high-pressure oil is routed to the center groove. All 15 ports in the manifold and carrier pistons are fed.

Middle reverse is accomplished when oil routes to the 15 pressure ports through the center groove, but also routes high-pressure oil to the inner groove. This forces some of the pistons to only recirculate high-pressure oil. The power stroke down, against the ramp, is equalled on the return stroke by the ramp forcing the piston against the high-pressure oil. When the power stroke is equalled by the return stroke, there is no power gain. The axle will turn slightly less than one and a half revolutions, using the same volume of oil as the low speed motor, because of greater friction.

High speed reverse is not possible because the internal friction of recirculating high-pressure oil by the majority of the pistons is greater than the functioning pistons can overcome.

OPERATION OF BRAKE

The brake assembly (Fig. 19) is used as a parking brake. It is activated by pulling the actuating rod, which causes the ball to expand the separator plates. As the separator plates expand, the brake disks are forced into contact with the separator plates, the bearing quill face and the brake housing, thus preventing rotation of the motor axles.

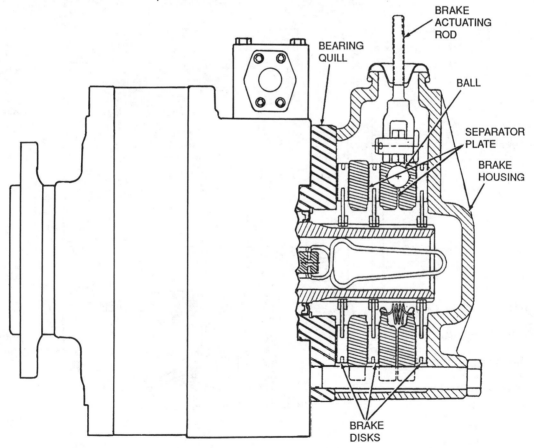

Fig. 19 — Brake Operation

OPERATION OF DESTROKE PUMP

Most cam lobe motors will have a destroke pump (Fig. 20). The destroke pump pressurizes the outer case when the motor is disengaged. Destroke pressure oil forces pistons in so that they will clear the cam lobes.

An eccentric washer rotates with the axle, activating the destoke pump once every revolution. The destroke pump draws low pressure oil from the inner case. Oil moves through the inlet check valve to the outer case.

The outer check valve regulates the oil pressure in the outer case. When oil pressure in the outer case reaches 70 kPa (10 psi), over the pressure in the inner case, the outlet check valve opens in order to maintain the pressure differential at that level. Oil from the outlet check valve can flow either to the inner case or out the drain line to the reservoir.

ADVANTAGES OF HYDROSTATIC DRIVES

- *Variable speeds and torques*
- *Easy one-lever control*
- *Smooth shifting without "steps"*
- *Shifts "on-the-go"*
- *High torque available for starting up*
- *Flexible location—no drive lines*
- *Compact size for big power*
- *Eliminates clutches and large gear trains*
- *Reduces shock loads*
- *Low maintenance and service*

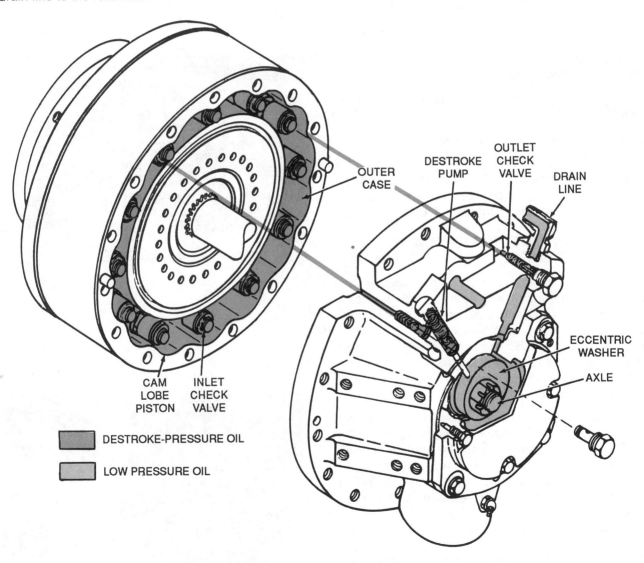

Fig. 20 — Destroke Pump Operation

SUMMARY OF MOTOR TYPES

This concludes our description of the three basic types of motors and the cam lobe motor.

Before going into the application and efficiency of these motors, let's review some of the points we have just covered.

In Summary:

1. A hydraulic motor is the opposite of a pump: A pump drives its fluid, while a motor is driven by its fluid.

2. A pump converts mechanical force into fluid force, while a motor converts fluid force into mechanical force.

3. A pump-motor drive works as follows: The pump is driven mechanically, drawing in fluid and pumping it to the motor. The motor is driven by the fluid from the pump and so drives its load by a mechanical link.

4. A motor is often quite similar to a pump in appearance and construction.

5. However, a hydraulic pump is really an actuator, like a cylinder.

6. The three basic types of motors are: gear, vane, and piston. All three are rotary in operation.

7. We have covered only the basic types of motors as well as one variation, the cam lobe motor. In actual use, there are many variations for special needs.

HYDRAULIC MOTOR APPLICATION AND EFFICIENCY

The first part of this chapter has described the physical construction and operation of the basic types of hydraulic motors as used on modern farm and industrial equipment. Now we must tell another part of the hydraulic motor story—how the hydraulic motor is used, why it is used, and how the three types of motors are rated in regard to power output, efficiency, size, etc.

Again, because of the wide variety of motors and hydraulic systems, we will not attempt to prescribe a particular motor for a particular application. We can only describe in a general way the good and bad points of each type, then let you judge why a particular type of motor is selected for a certain application.

MOTOR SELECTION

To select a hydraulic motor, we must first find out what we expect to achieve by using the motor. This, of course, means analyzing the system's require-

Fig. 21 — A Hydrostatic Drive Propels This Grain Combine

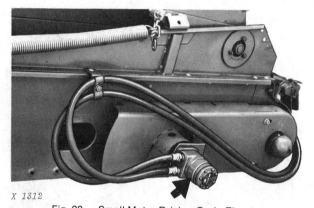

Fig. 22 — Small Motor Driving Grain Elevator

Fig. 23 — Large Motor Driving Self-Propelled Machine

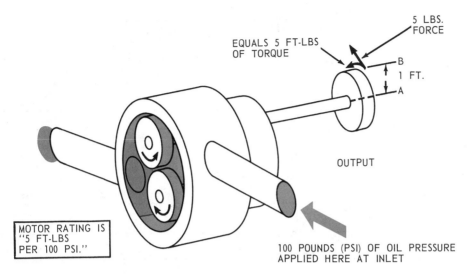

EQUALS 5 FT-LBS
OF TORQUE

5 LBS.
FORCE

B

1 FT.

A

BUT IF SYSTEM
APPLIED 1000 PSI
AT INLET, MOTOR
TORQUE WOULD
EQUAL 50 FT-LBS
AT POINT B.

OUTPUT

MOTOR RATING IS
"5 FT-LBS
PER 100 PSI."

100 POUNDS (PSI) OF OIL PRESSURE
APPLIED HERE AT INLET

X3404

Fig. 24 — What a Motor Torque Rating Means

ments, then selecting a motor that best meets these needs. This does not mean that a highly rated motor in terms of power, displacement, efficiency, etc. is best for every application, nor does it mean that using a motor that barely meets the system requirements is good practice, either. It does mean that each application dictates the type of motor to be used, regardless of how it is rated or compared with the other types.

MOTOR TORQUE

The first consideration is to have a motor which will turn the load. However, torque is directly related to the input oil pressure. For this reason, most motor torque ratings are given as *torque per 100 psi input.*

Torque should always be calculated at maximum loads. The torque needed to start a load is always greater than the torque needed to maintain rotation. So the *starting torque* should be used in selecting a motor.

Example: We have a motor rated "5 foot-pounds per 100 psi." This means that if we have 100 psi of system pressure (see Fig. 24), we can exert a 5 pound force one foot from the shaft center (point B). If our system pressure were 1000 psi, our maximum torque would be 50 pounds force at point B or *50 foot-pounds of torque* in the shaft.

As you have seen in the example, torque is a force applied at a *distance from the center of a shaft.* It

is expressed in terms of inches and feet (5 foot-pounds equals 60 inch-pounds).

In the actual selection of a motor, you must reverse the procedure.

Example: Our system pressure is 1500 psi, and our maximum load is 50 ft.-lbs. of torque. Our requirement will then be for a motor with a torque rating of *3 ft.-lbs. of torque per 100 psi of input.*

You will note that torque is affected by oil *pressure* only. Volume of oil does not change torque.

SPEED

Having determined the torque requirements of the motor, you must then supply enough oil to achieve the proper speed. With a given supply pump, only the *volume* of oil affects *speed.*

MOTOR HORSEPOWER

Horsepower is the measure of total work done by the motor. It is the combination of *force × speed,* one horsepower being the work done in moving 33,000 pounds a distance of 1 foot in a time of 1 minute.

Motor horsepower is the most common method of rating hydraulic motors.

MOTOR SIZE

One of these practical factors is the physical size of the hydraulic motor. As you well know, space for hydraulic components is limited on most mobile

MOTOR COMPARISON CHART

| | GEAR MOTORS | | VANE MOTORS (Balanced) | PISTON (AXIAL) MOTORS | | CAM LOBE MOTORS |
	External	Internal		Fixed Displ.	Var. Displ.	
Physical Size	Small to Medium	Small to Medium	Small to Medium	Medium to Large	Medium to Large	Medium to Large
Average Weight to Power Ratio-lb/hp	0.9	0.9	1.0	1.4	3.2	5-13
Pressure Range (psi)	100-2000	100-2000	100-2500	100-5000+	100-5000+	100-5000+
Speed Range (rpm)	100-3000	100-5000	10-3000	10-3000	10-3000	1-220
Actual Torque (% of Theoretical)	80-85	80-85	85-95	90-95	90-95	90-95
Starting Torque (% of Theoretical)	70-80	75-85	75-90	85-95	85-95	85-95
Momentary Overload Torque (% of Actual)	110-120	115-130	120-140	120-140	120-140	130-150
Volumetric Efficiency (%)	80-90	85-90	85-90	95-98	95-98	95-100
Over-all Efficiency (%)	60-90	60-90	75-90	85-95	85-95	85-95
Estimated Bearing Life (hrs.) @ 1/2 load	5000-10000	5000-10000	7000-15000	15000-25000	15000-25000	20000-25000
Displacement	Fixed	Fixed	Fixed	Fixed	Variable	Variable in Steps
Reversibility	Possible	Possible	Possible	Very Good	Very Good	Good at Full Displacement
How Operates as a Pump	Good	Good	Good	Very Good	Very Good	Very Good
Estimated Bearing Life (hrs.) @ full load	2000-5000	2000-5000	3000-6000	7000-15000	7000-15000	3000-5000

NOTE: Remember that the values in this chart are not absolute. They may vary with each particular model of motor.

equipment. Fortunately, the three basic types of motors used on mobile equipment come in a wide range of sizes. However, some problems may be experienced if the motor must be of a larger size to provide the necessary power. In such cases, power and other system requirements would take precedence over motor size.

MOTOR PRESSURE RANGE

Aside from the fact that pressure affects torque and horsepower, you should know the pressure limits when selecting a motor. Never install a motor that cannot accept the full pressure range of the system; this would require extra valving to control pressure or could damage the motor. By the same token, a motor with a greater pressure range than the system would be a waste of money.

MOTOR SPEED RANGE

Speed is another important factor. To require a motor to operate at speeds lower than recommended is wasteful because of increased slippage and leakage losses. Speed also affects horsepower output so care must be taken to insure that motor speed is capable of producing the necessary horsepower.

TORQUE—RUNNING, STALLED AND OVERLOAD

Since most motors are rated in terms of theoretical torque, the percentages of actual, stalled, and momentary overload torque are important. For instance, a motor with a high theoretical torque may have a low actual running torque or a low starting or stalled torque and may not be able to handle a large load. At some time during operation, the motor may also be called upon to deliver more torque than its normal running torque for a short period of time.

MOTOR EVALUATION

Now that we have given you some ideas as to how a motor is evaluated, let's see how the three basic motors compare with one another.

The chart above compares the three motors in a very general sense. Remember, we are not favoring any particular type of motor or any particular brand of motor. We are simply relating to you what is generally known about the three types of motors. There are cases when a type of motor may be rated above what we have indicated here. For this reason, we suggest that you thoroughly investigate the many brands of motors available before selecting a particular motor.

MISCELLANEOUS FACTORS

Other factors which have a bearing on motor selection and identification are: running volumetric efficiency, over-all efficiency, leakage at stalled torque, bearing life, displacement (fixed or variable), reversability, ability to act as a pump as well as a motor, weight- or size-to-power ratio, ease of maintenance, and initial cost. All of these values play some part in selecting and judging a motor.

HYDRAULIC MOTOR MALFUNCTIONS

In this chapter, we've shown the great similarity between the motor and the pump.

This similarity extends into the realm of motor failures as well.

The majority of motor problems fall into these categories:

- **Improper Fluid**
- **Poor Maintenance**
- **Improper Operation**
- **Poor Motor Selection**
- **Improper System Design**
- **Mechanical Failures**

Let's discuss each one.

IMPROPER FLUID

The motor is no different than any of the other components of the hydraulic system—it must have clean fluid, in adequate supply, and of the proper quality and viscosity.

Failures due to improper fluid are covered quite extensively in Chapters 2 and 10 and much of this information is applicable to hydraulic motors. For this reason, we won't go into the subject here, but instead refer you to those chapters.

POOR MAINTENANCE

A poor maintenance program runs a close second in the list of major problems. We find many causes of failures in this category, all of which can be attributed to human error. Some of these causes are:

1) Failure to check and repair lines and connections for air or fluid leaks. This can allow dirt and air into the system, plus lowering pressure and causing erratic operation.

2) Failure to check and repair other components such as pumps, control valves, or filters. This can lead to excessive wear of internal parts due to excessively high or low pressure, excessive or inadequate fluid.

3) Failure to install the motor correctly. Motor shaft misalignment can cause bearing wear which can lead to loss of efficiency. A misaligned shaft can also reduce torque, increase friction drag and heating, and result in shaft failure.

4) Failure to inspect fluid for proper quality and quantity (see Chapter 10).

5) Failure to find the cause of a motor malfunction. If the motor does fail, always look for the *cause* of the failure first. It is obvious that if the cause is not corrected, the failure will recur.

Maintenance problems are covered in more detail in Chapter 11 of this manual.

IMPROPER OPERATION

Exceeding the operating limits is a prime way to promote motor failure. Every motor has certain design limitations on pressure, speed, torque, displacement, load, and temperature. These limits, of course, vary from motor to motor and are set up by the manufacturer.

The list below describes what can happen if these limits are exceeded:

Excessive Pressure—can cause wear due to lack of lubrication, generate heat due to motor slippage, or cause motor to exceed torque limits.

Excessive Speed—can cause heating due to slippage, or cause wear of bearings and internal parts.

Excessive Torque—can cause fatigue and stress to bearings and shaft, especially on applications that require frequent motor reversing.

Excessive Displacement—can generate heat due to pressure across motor not producing usable work.

Excessive Load—can create bearing and shaft fatigue.

Excessive Temperature—can cause loss of efficiency and speed due to thinned oil, and can produce rapid wear due to lack of lubrication.

Cold weather starting is another common abuse of the motor. Most motor manufacturers give detailed instructions on how to start their motors. Failure to follow these instructions can result in wear and galling, due to lack of lubrication; torque loss below stall torque specifications; and stress and fatigue on bearings and shaft.

POOR MOTOR SELECTION, IMPROPER SYSTEM DESIGN, MECHANICAL FAILURES

The last three problems are obvious. They are not big problems, but do deserve attention to prevent them from becoming so.

Poor motor selection has been covered in the motor evaluation section which tells the importance of selecting a motor that meets the requirements of the application.

A properly designed system insures that all lines are of the right size and that they have no sharp bends to create friction and fluid overheating. It also insures that the correct valving is used to control the operation of both the pump and the motor. And lastly, it guarantees that the motor and the system are compatible.

Mechanical failures due to normal wear and operation are not preventable. But they can be anticipated by keeping accurate maintenance and operating records and by using a good preventive maintenance program (see Chapter 11).

SUMMARY: MOTOR APPLICATION AND EFFICIENCY

To sum up this section, here is what must be done to get good operation and long life from the motor.

1. Follow the manufacturers' recommendations and specifications to the letter. This means operating within the motor's limits and repairing the motor with only authorized parts.

2. Select a motor that meets all the requirements of the system and the application. This doesn't mean that the motor should exceed these requirements—often that can cause poor operation and loss of efficiency, too.

3. Have a preventive maintenance program, complete with accurate maintenance and operating records.

CONCLUSION: THE USES OF HYDRAULIC MOTORS

In concluding this chapter, there are several points about hydraulic motors that should be emphasized.

Hydraulic motors are becoming more prominent in systems or applications that had previously used electrical motors or reciprocating engines as the source of power. There are two reasons for this:

1. The use of hydraulic power is becoming more widespread, especially on mobile agricultural and industrial equipment.

2. The hydraulic motor, when fluid power is available, has many advantages over the other sources of power.

The greatest advantage of the motor is the size or weight-to-power ratio. In comparing an electrical motor with a hydraulic motor of the same horsepower, you will find that the hydraulic motor is several times smaller—a desirable quality on mobile equipment.

Another advantage is the ease of control. Very little energy is expended to supply a great deal of force. A good example of this is in power steering.

Still another advantage is accuracy of control. Smooth steady power in small or large quantities is ready at the touch of the hand. No belts or chains to delay the time between actuating controls and the power source.

Simplicity, economy, and safety round out the list. All of these factors are gained by using fewer moving parts in the motor.

In summary, the hydraulic motor can be a prime source of power when it meets the application requirements, is operated according to specifications, and is properly serviced and maintained.

NOTE: For more information on hydrostatic drives, refer to *Chapter 5* in *F.O.S. Power Trains*.

DIAGNOSING MOTOR FAILURES

The following trouble shooting charts are given as a general guide to some of the common motor failures, what the cause might be, and the possible remedy.

I. MOTOR WON'T TURN

Possible Cause	Possible Remedy
1. Shaft seizure due to: a. Excessive loads.	a. Check load and load capacity of the motor.
b. Lack of lubrication.	b. Check fluid level and quality Check operating pressure and temperature.
c. Misalignment.	c. Correctly align shaft with work load.
2. Broken shaft.	2. Replace shaft. Find reason for breakage.
3. No incoming pressure.	3. Check and repair clogged, leaking, or broken lines or passages.
4. Contaminated fluid.	4. Check and clean entire hydraulic system (Chapter 11). Find source of contamination. Install clean fluid of proper quantity and quality.

II. SLOW MOTOR OPERATION

1. Wrong fluid viscosity.	1. Install clean fluid of proper quantity and quality.
2. Worn pump or motor.	2. Check pump and motor specifications. Repair or replace if necessary.
3. High fluid temperature.	3. Check for line restrictions, wrong fluid viscosity, or low fluid level.
4. Plugged filter.	4. Check cause of plugging, and clean or replace filter.

III. ERRATIC MOTOR OPERATION

1. Low pressure.	1. Check for air or fluid leaks.
2. Inadequate fluid flow.	2. Check for air or fluid leaks. Check system capacity.
3. System controls failing.	3. Check pump and control valves for proper operation.

IV. MOTOR TURNS IN WRONG DIRECTION

1. Pump-to-motor connections wrong.	1. Reconnect pump and motor.
2. Wrong timing.	2. Check manufacturer's specifications.

V. MOTOR SHAFT NOT TURNING

1. Excessive work loads.	1. Check motor load specifications.
2. Internal motor parts worn.	2. Replace parts or complete motor as necessary.

TEST YOURSELF
QUESTIONS

1. (Fill in the blanks.) A hydraulic motor converts _____ force into _____ force.

2. How is a motor different in operation from a pump?

3. What three types of motors are most popular on modern farm and industrial systems?

4. What is the "torque" of a motor?

5. (True or false?) "A pump can generally be used as a motor."

6. What are the two types of cam lobe motors?

6—1

HYDRAULIC ACCUMULATORS / CHAPTER 6

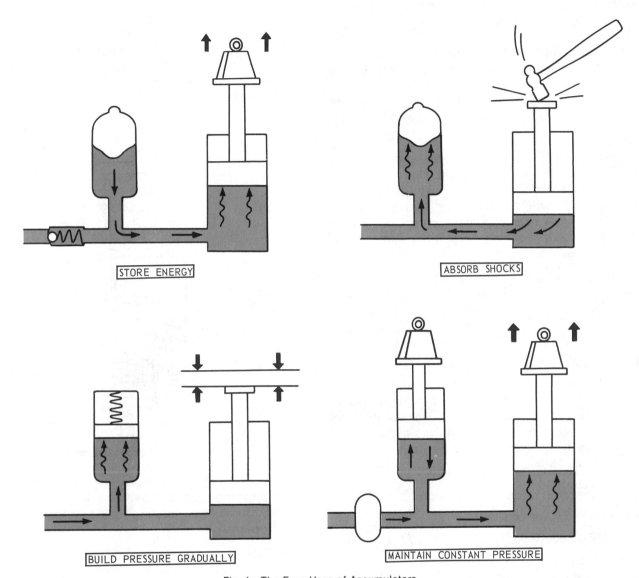

STORE ENERGY

ABSORB SHOCKS

BUILD PRESSURE GRADUALLY

MAINTAIN CONSTANT PRESSURE

X 1262

Fig. 1—The Four Uses of Accumulators

> ⚠️ **CAUTION: Accumulators store energy. Before working on a hydraulic system with an accumulator, you must relieve all pressure.**

A spring is the simplest accumulator. When compressed, a spring becomes a source of potential energy. It can also be used to absorb shocks or to control the force on a load. Hydraulic accumulators work in much the same way. Basically they are containers which store fluid under pressure.

USES OF ACCUMULATORS

Accumulators have four major uses (Fig. 1):

- **Store Energy**
- **Absorb Shocks**
- **Build Pressure Gradually**
- **Maintain Constant Pressure**

While most accumulators can do any of these things, their use in a system is usually limited to only one.

Accumulators which STORE ENERGY are often used as "boosters" for systems with fixed displacement pumps. The accumulator stores pressure oil during slack periods and feeds it back into the system during peak periods of oil usage. The pump recharges the accumulator after each peak. Sometimes the accumulator is used as a protection against failure of the oil supply. Example: power brakes on larger machines. If the system oil supply fails, the accumulator feeds in several "charges" of oil for use in emergency braking.

Accumulators which ABSORB SHOCKS take in excess oil during peak pressures and let it out again after the "surge" is past. This reduces vibrations and noise in the system. The accumulator may also smooth out operation during pressure delays, as when a variable displacement pump goes into stroke. By discharging at this moment, the accumulator "takes up the slack."

Accumulators which BUILD PRESSURE GRADUALLY are used to "soften" the working stroke of a piston against a fixed load, as in a hydraulic press. By absorbing some of the rising oil pressure the accumulator slows down the stroke.

Accumulators which MAINTAIN CONSTANT PRESSURE are always weight-loaded types which place a fixed force on the oil in a closed circuit. Whether the volume of oil changes from leakage or from heat expansion or contraction, this accumulator keeps the same gravity pressure on the system.

TYPES OF ACCUMULATORS

The major types of accumulators are:

- **Pneumatic (Gas-Loaded)**
- **Weight-Loaded**
- **Spring-Loaded**

PNEUMATIC ACCUMULATORS

We learned in Chapter 1 that fluids will not compress, but gases will. For this reason, many accumulators use inert gas as a way of "charging" a load of oil or of providing a "cushion" against shocks. An inert gas is a gas that will not explode.

"Pneumatic" means operated by compressed gas. In these accumulators, gas and oil occupy the same container. When the oil pressure rises, incoming oil compresses the gas. When oil pressure drops, the gas expands, forcing out oil.

In most cases, the gas is separated from the oil by a piston, a bladder, or a diaphragm. This prevents mixing of the gas and oil and keeps gas out of the hydraulic system.

(Some accumulators for low-pressure or fairly static uses do not have a separator between the gas and oil, but their use is very limited in modern hydraulics. Also, there is no way to precharge these accumulators.)

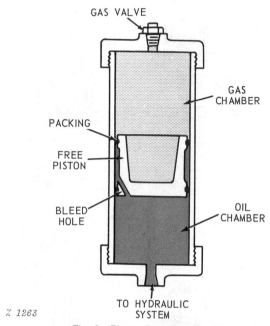

Fig. 2—Piston-Type Accumulator

A typical PISTON-TYPE ACCUMULATOR is shown in Fig. 2. It looks like a hydraulic cylinder minus the piston rod. A "free-floating" piston separates the gas from the oil.

The piston fits into a smooth bore and uses packings to separate the gas from the oil. With double packings, a bleed hole is needed as shown to relieve pressure of any oil seeping into the center area between the two packings.

The accumulator can be "precharged" with gas before use in a system. This is done by filling the gas chamber to a desired pressure with an inert gas such as dry nitrogen.

Piston-type accumulators require careful service to prevent leakage. But they offer a high power output for their size and are very accurate in operation.

Fig. 3—Bladder-Type Accumulator

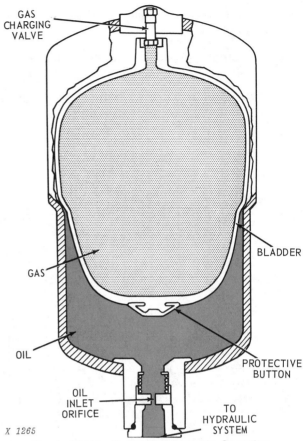

X 1265

Fig. 4—Bladder-Type Accumulator in Operation

BLADDER-TYPE ACCUMULATORS are shown in Figs. 3 and 4. A flexible bag or bladder made of synthetic rubber contains the gas and separates it from the hydraulic oil. The bladder is molded to the gas charging stem located at the top of the accumulator.

To prevent damage to the bladder, a protective button is used at the bottom (Fig. 4). This button prevents the bladder from being drawn into the oil port when the bladder expands. Otherwise, the bladder might be cut or torn.

In the accumulator shown in Fig. 4, the spring-loaded orifice admits a free flow of oil, but meters oil coming out for a slower, gauged action.

Bladder-type accumulators can also be precharged before use.

DIAPHRAGM-TYPE ACCUMULATORS use a metal element to separate the gas from the oil. Molded to the element is a rubber diaphragm which flexes in response to pressure changes (Fig. 5). These accumulators are light in weight and are often used in aircraft systems.

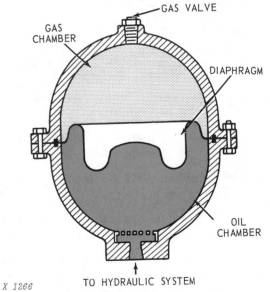

X 1266

Fig. 5—Diaphragm-Type Accumulator

Effects of Different Precharges on Pneumatic Accumulators

When you precharge an accumulator with gas, how much pressure should you use?

It depends upon how you want the accumulator to work.

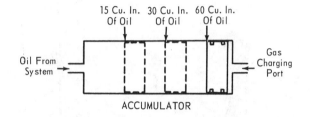

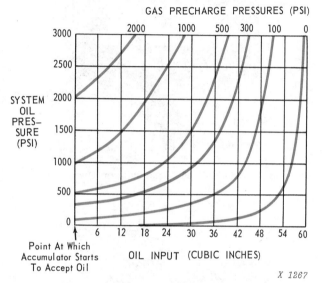

Fig. 6—Effect of Different Precharges on a Pneumatic Accumulator

Fig. 6 shows how six different precharges effect the operation of an accumulator.

The gas precharges shown are 2000, 1000, 500, 300, 100 and 0 psi (see top line).

The accumulator has a 60 cubic inch displacement for input oil (bottom line).

As system oil pressure rises (vertical scale at left), incoming oil displaces gas in the accumulator.

But the accumulator with the highest gas precharge (2000 psi) starts to accept oil much later than the ones which have lower precharges. In other words, system oil pressure must rise above 2000 psi to even start to load the accumulator with the highest precharge.

The precharge also effects *how much* oil the accumulator will accept at a given pressure.

The lower the precharge, the more oil the accumulator will load up at a given pressure.

In summary, the selection and use of a pneumatic accumulator depends upon the pressure and volume needs of the system. In other words, what pressures are required and their limits, and what volume of oil is needed to dissipate oil or supply it to the system.

Precautions for Pneumatic Accumulators

Observe the following precautions when working on pneumatic accumulators. The correct procedures for service are given in detail later under "Servicing and Precharging Pneumatic Accumulators."

⚠ **1. CAUTION: NEVER FILL AN ACCUMULATOR WITH OXYGEN! An explosion could result if oil and oxygen mix under pressure.**

2. Never fill an accumulator with air. When air is compressed, water vapor in the air condenses and can cause rust. This in turn may damage seals and ruin the accumulator. Also, once air leaks into the oil, the oil becomes oxidized and breaks down.

3. Always fill an accumulator with an *inert gas* such as *dry nitrogen*. This gas is free of both water vapor and oxygen; this makes it harmless to parts and safe to use.

4. Never charge an accumulator to a pressure more than that recommended by the manufacturer. Read the label and observe the "working pressure."

5. Before removing an accumulator from a hydraulic system, *release all hydraulic pressure*.

6. Before you disassemble an accumulator, *release both gas and hydraulic pressures*.

7. When you disassemble an accumulator, make sure that dirt and abrasive material does not enter any of the openings.

Servicing and Precharging Pneumatic Accumulators

CHECKING PRECHARGED ACCUMULATOR ON THE MACHINE

1. If you suspect external gas leaks, apply soapy water to the gas valve and seams on the tank at the "gas" end. If bubbles form, there is a leak.

2. If you suspect internal leaks, check for foaming oil in the system reservoir and/or no action of the accumulator. These signs usually mean a faulty bladder or piston seals inside the accumulator.

3. If the accumulator appears to be in good condition but is still slow or inactive, precharge it as necessary (see below).

BEFORE REMOVING ACCUMULATOR FROM MACHINE

First be sure all hydraulic pressure is released. To do this, shut down the pump and cycle some mechanism in the accumulator hydraulic circuit to relieve oil pressure (or open a bleed screw).

REMOVING ACCUMULATOR FROM MACHINE

After all hydraulic pressure has been released, remove the accumulator from the machine for service.

REPAIRING ACCUMULATOR

1. *Before dismantling accumulator, release all gas pressure.* Normally unscrew the gas valve lever very slowly. Install the charging valve first if necessary. Never release the gas by depressing the valve core, as the core might be ruptured.

2. Disassemble the accumulator on a *clean* bench area.

3. Check all parts for leaks or other damage.

4. Plug the openings with plastic plugs or clean towels as soon as parts are removed.

5. Check bladder or piston seals for damage and replace if necessary.

6. If gas valve cores are replaced, be sure to use the recommended types.

7. Carefully assemble the accumulator.

 CAUTION: Incorrect charging procedures can be dangerous. Only charge the accumulator yourself if you have the know-how and equipment to do so safely. If in doubt, have it charged by a professional.

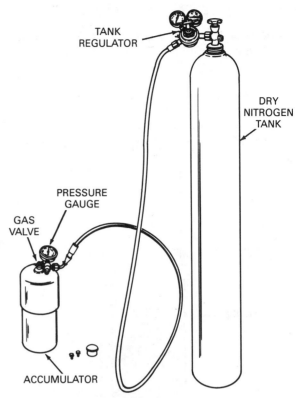

Fig. 7—Precharging Accumulator (Pneumatic Type)

PRECHARGING ACCUMULATOR

Attach the hose from a *Dry Nitrogen* tank to the charging valve of the accumulator and open the accumulator charging valve (Fig. 7).

Open the valve on the regulator very slowly until pressure on gauge is same as that recommended by the manufacturer. Close the charging valve on the accumulator, then close the valve on the regulator. Remove hose from charging valve.

NOTE: When checking precharge on an accumulator installed on a machine, first release hydraulic pressure from the accumulator. Otherwise you will not get a true pressure reading.

INSTALLING ACCUMULATOR ON MACHINE

Attach accumulator to machine and connect all lines. Start machine and cycle a hydraulic function to bleed any air from the system. Then check the accumulator for proper action.

WEIGHT-LOADED ACCUMULATORS

The earliest form of accumulator is the weight-loaded type (Fig. 8).

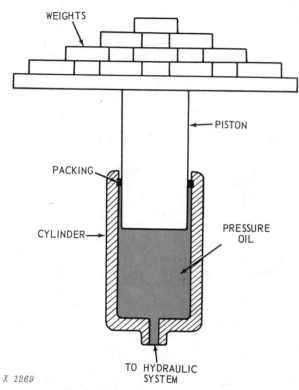

Fig. 8—Weight-Loaded Accumulator

This accumulator uses a piston and cylinder, but heavy weights on the piston do the job of loading or charging the oil. It is loaded by gravity.

Operation is very simple. The pressure oil in the hydraulic circuit is pushed into the lower oil chamber. This raises the piston and weights. The accumulator is now charged, ready for work. When oil is needed, pressure drops in the system and gravity forces the weights and piston down, discharging the oil into the system.

The advantage of the weight-loaded accumulator is that it can provide constant pressure.

The disadvantages are the bulky size and heavy weight. For mobile machines, there is little use for the gravity accumulator.

SPRING-LOADED ACCUMULATORS

This accumulator is very similar to the weight-loaded type except that springs do the loading.

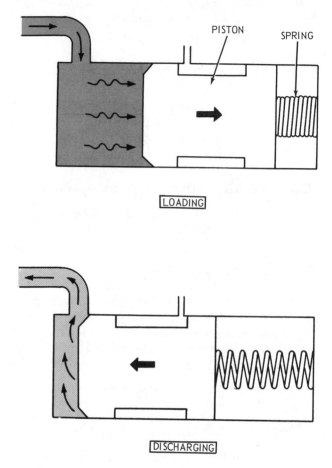

Fig. 9—Spring-Loaded Accumulator in Operation

In operation, pressure oil loads the piston by compressing the spring (Fig. 9). When pressure drops, the spring forces oil into the system.

The accumulator shown in Fig. 9 is an internal type used as a gradual pressure builder for an automatic transmission. When the transmission is shifted, pressure drops and the accumulator sends a "surge" of oil in to "take up slack." This fills the chamber behind the clutch pistons. Then pressure builds gradually for a smooth engagement of the clutch.

By controlling the flow of oil to the accumulator, the time needed to charge it can also be controlled (Fig. 10). This is commonly done where a "cushioned" engagement of hydraulic components is needed. In the graph, note how pressure drops as the accumulator discharges, then builds slowly as it recharges, rising sharply to full pressure again when it refills with oil. The time lag during the charging cycle can be speeded up or slowed down by feeding in more or less oil to the accumulator.

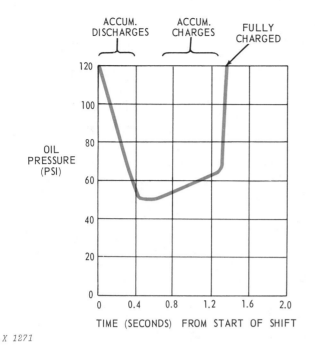

X 1271

Fig. 10—Operating Cycle of a Spring-Loaded Accumulator

The operation of spring-loaded accumulators can be varied by changing 1) the strength of the spring, 2) the length of the spring, 3) the preload on the spring, 4) the size of the piston or, 5) the length of the piston stroke.

Fig. 11 shows the effect of using a stronger spring with less preload (B) as opposed to a weaker spring (A).

Now we can see how important the correct spring is in the operation of these accumulators.

Be sure to use genuine parts and to follow the manufacturer's recommendations when repairing accumulators.

In another design shown in Fig. 12, a disk on the end of the piston mounts several rods which hold the springs. As oil pressure raises the piston, the springs are compressed against the solid shoulder of the cylinder. The springs are preset by tightening the adjusting nuts.

The advantage of spring-loaded accumulators is that they never have to be precharged or recharged.

The disadvantage is that these accumulators are too bulky when designed for high volume or high pressure systems. Therefore, they are practical only for low volume or low pressure uses.

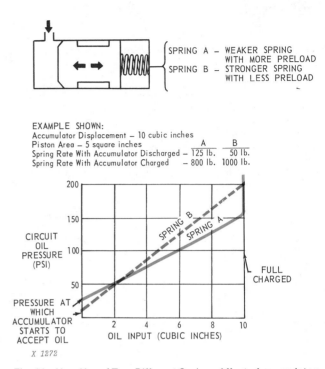

EXAMPLE SHOWN:
Accumulator Displacement – 10 cubic inches
Piston Area – 5 square inches

	A	B
Spring Rate With Accumulator Discharged –	125 lb.	50 lb.
Spring Rate With Accumulator Charged –	800 lb.	1000 lb.

X 1272

Fig. 11—How Use of Two Different Springs Affects Accumulator Operation

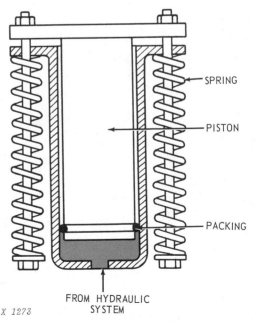

X 1273

Fig. 12—Spring-Loaded Accumulator

TEST YOURSELF

QUESTIONS

1. Why is oxygen NOT recommended for use in a pneumatic accumulator?

2. Why is air not recommended for use in a pneumatic accumulator?

3. What should you do first of all before removing an accumulator from a machine?

4. What are the four major uses of accumulators?

HYDRAULIC FILTERS / CHAPTER 7

WHY ARE FILTERS USED?

Did you ever stop to think that hydraulic fluids are lubricants for precision parts as well as a means of transmitting power?

Contaminated oil can score or completely freeze a precisely fitted valve spool. Dirty oil can ruin the close tolerance of finely finished surfaces, and a grain of sand in a tiny control orifice can put a whole machine out of operation. It's not hard to see that you have to keep the oil clean if you want a hydraulic system to operate without trouble.

Fig. 1—Dust is a Major Source of Contamination

It's a constant battle because dirt is everywhere. Look at Fig. 1 and you see that the air surrounding a machine is a major source of contamination.

Another source of contaminants is the machine itself. As it works and wears normally, the machine generates burrs and chips of metal.

Measured in dollars and cents, it's a whole lot cheaper to buy a good filter, to maintain it properly, and to keep your oil clean than it is to replace a pump or a valve that is worn out by contamination.

HOW FILTERS ARE USED

A FULL-FLOW system filters the entire supply of oil each time it circulates in the hydraulic system. Filters in a full-flow system are usually located in the pump inlet line and in the return line to the reservoir. Additional filters, of course, may be located in front of or behind other hydraulic components if they are needed.

In contrast, a BYPASS filter system has its filter connected to a tee in the pressure line so that only a small portion of each oil cycle is diverted through the filter. The remainder of the oil goes unfiltered to the system or to the reservoir.

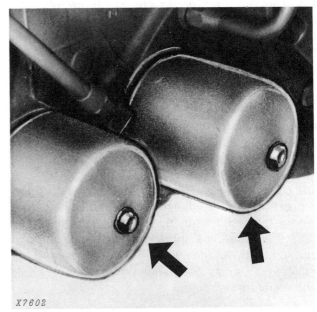

X7602

Fig. 2—Tractor Transmission—Hydraulic System Filter

The location of the filter in a hydraulic system will vary with the design of the machine. Fig. 2 shows a filter that is an integral part of the machine, while Fig. 3 shows a filter hooked into an outside line. Regardless of location, the one purpose of the filter is to keep the oil clean.

Fig. 3—Combine Hydraulic Drive System Filter

Filtration occurs as oil passes through the filter. Fig. 4 shows a full-flow system with the inlet and return oil filters arranged in one package. Oil from the reservoir enters the inlet screen and, after being filtered, flows to the pump. Then the oil is pumped to the control valve and cylinders and is filtered a second time as it passes through the return filter to the reservoir.

A small pressure difference exists between the inside and outside of even a new filter because the oil is being restricted as it passes through—like pushing a screen door open in a high wind. If you cover the screen and then try to open the door, a great deal of resistance is felt. The same is true with a hydraulic filter. As the filter gets dirty, the pressure difference increases, finally to a point where no oil will flow when the filter is completely plugged. This may also happen when oil is very cold and therefore less fluid.

To prevent pressure from building up so high that it might break the filter or starve one of the hydraulic components, a relief valve is usually used to bypass oil around the filter. (Don't confuse this with a bypass filter system.)

Fig. 5 shows such a relief valve in operation. Please note, however, that the inlet screen has no relief valve protection. Because of the difference in degree of filtration, the return filter element will plug much sooner than the inlet screen.

Of course when the relief valve opens, dirty oil pours into the hydraulic system. Unless the filters are serviced immediately, the dirt in the oil will increase wear in hydraulic components and the inlet screen will continue to plug until pump starvation occurs.

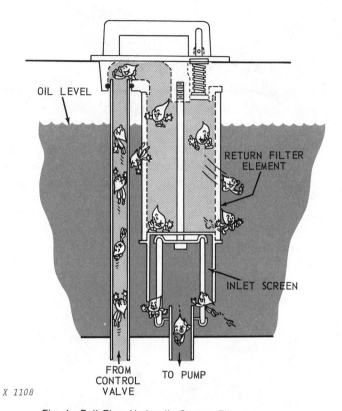

Fig. 4—Full-Flow Hydraulic System Filters

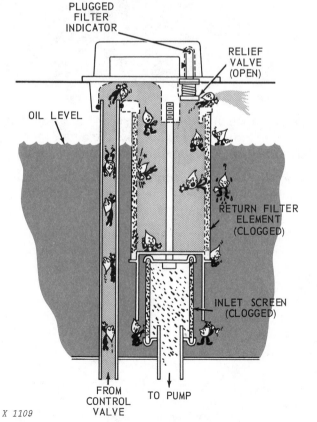

Fig. 5—Hydraulic Filter Relief Valve

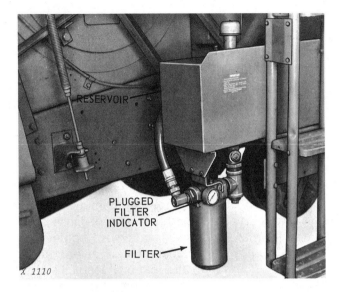

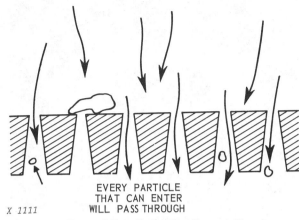

X 1111

Fig. 7—Tapered Flow Path Helps Prevent Plugging

The indicators for plugged filters shown in Figs. 5 and 6 make it easy to see when the filters require servicing.

With a relief valve in a filter, it's easy to see how important it is to use the right filter and hydraulic oil. If the wrong filter is used, or oil that is too heavy is put in the reservoir, the pressure difference between the inside and outside of a filter can be so great that it exceeds the relief valve setting. When this happens, the valve will open and the oil will never be filtered. Newer machines may have a "plugged filter" indicator that alerts to the need for filter service.

TYPES OF FILTERS

Now let's look at the type of filters used in a hydraulic system and just how much filtering they actually do.

Filters can be classified as either surface-type filters or depth-type filters depending on the way they remove dirt from hydraulic oil.

SURFACE FILTERS have a single surface that catches and removes dirt particles larger than the holes in the filter. Dirt is strained or sheared from the oil and stopped outside the filter as oil passes through the holes in a straight path (Fig. 7). Many of the large particles will fall to the bottom of the reservoir or filter container, but eventually enough particles will wedge in the holes of the filter to prevent further filtration. Then the filter must be cleaned or replaced.

X 1112

Fig. 8—Wire Mesh Filter

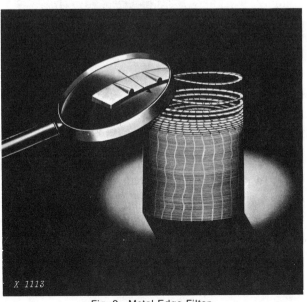

X 1113

Fig. 9—Metal Edge Filter

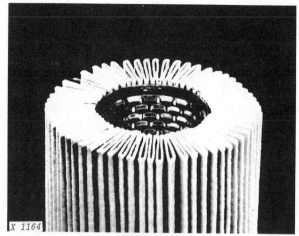

X 1164

Fig. 10—Pleated Paper Filter

A surface filter may be made of fine wire mesh (Fig. 8), stacked metal or paper disks, metal ribbon wound edgewise to form a cylinder (Fig. 9), cellulose material molded to the shape of the filter, or accordian-pleated paper (Fig. 10).

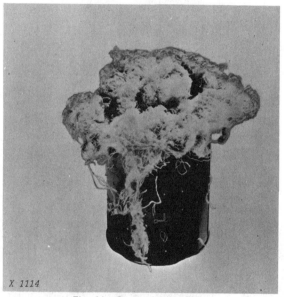

X 1114

Fig. 11—Cotton Waste Filter

DEPTH FILTERS, in contrast to the surface type, use a large volume of filter material to make the oil move in many different directions before it finally gets into the hydraulic system (Fig. 11).

Depth filters can be classified as either **absorbent** or **adsorbent,** depending on the way they remove dirt.

Absorbent filters operate mechanically like a sponge soaking up water. Oil passes through a large mass of porous material such as cotton waste, wood pulp, wool yarn, paper or quartz, leaving dirt trapped in the filter. This type of filter will remove particles suspended in the oil and some water and water soluble impurities.

Adsorbent filters operate the same way as absorbent filters but also are chemically treated to attract and remove contaminants. This filter may be made of charcoal, chemically-treated paper or fuller's earth. It will remove contaminating particles, water soluble impurities and, because of its chemical treatment, will also remove contamination caused by oil oxidation and deterioration. Adsorbent filters may also remove desirable additives from the oil and for this reason are not often used in hydraulic systems.

DEGREES OF FILTRATION

In addition to the type of filter, the degree of filtration is also important to a hydraulic system for it is the degree of filtration that tells just how small a particle the filter will remove. The most common measurement used to determine degree of filtration is a micron (one micron is approximately 0.00004-inch or 40 millionth of an inch). To get an idea of how small a micron really is, 25,000 particles of this size would have to be laid side by side to total just one inch.

The smallest particle that can normally be seen with an unaided eye is about 40 microns, so much of the dirt that is filtered out of a hydraulic system is invisible. For example, grain combines operating in the field today have filters that will remove particles as small as 10 microns in diameter or about one-tenth the size of a grain of table salt.

Some filters, such as those made of wire mesh, may allow particles as big as 150 microns to pass. Although they do not provide as fine a cleaning action as some other types of filters, wire mesh offers less resistance to oil flow and is often used on pump inlet lines to prevent the possibility of starvation.

A filter will pass small solids and stop larger ones but the actual amount of filtering done is difficult to determine. Because the material that is stopped by the filter is not taken away continuously, the size of the openings in the filter will usually decrease with use.

Two or more particles smaller than the holes in the filter may approach at the same time and become wedged in the hole. The result is that the hole will now remove dirt particles far smaller than it would when new. As more dirt wedges in the filter, the holes become smaller and eventually are plugged solid.

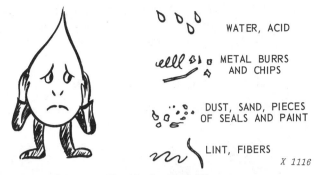

Fig. 13—Contamination

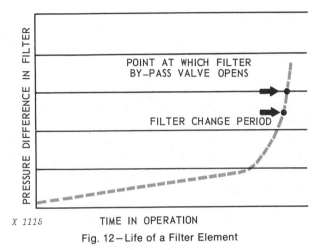

Fig. 12—Life of a Filter Element

Fig. 12 shows the gradual reduction in the size of the filter pores until, near the end of filter life, the pressure difference between the inside and outside of the filter rises sharply. At this point, the filter stops operating and should be cleaned or replaced.

CONTAMINATION

But just what is contamination and how does it get into a hydraulic system?

Liquids, metallic particles, non-metallic particles and fibers are all forms of material that can contaminate the oil (Fig. 13). This material can come from both inside and outside the hydraulic system.

The air is a prime source of contamination. It may contain moisture and particles from the atmosphere, as well as road or field dust. These contaminants can enter through breathers and filler pipes, past seals and gaskets, or when the system is opened for repair or maintenance.

The hydraulic machine itself is a significant source of contamination. During break-in, bits of metal and other abrasive particles will contaminate the oil. Later, during normal operation, the machine will generate its own contamination as fragments of paint, pieces of seals and gaskets, and metallic particles caused by wear begin to get into the oil.

Hydraulic oil can be contaminated during maintenance and service if unclean containers and funnels, dirty oil, or dirty and linty wiping cloths are used.

IMPORTANT: Always cap or plug open lines or connectors to reduce the possibility of contamination.

Fig. 14—Mishandling Will Ruin a Filter

Mishandling of filters can reduce their effectiveness and even cause pieces to break off and get into the oil (Fig. 14).

The oil itself is another source of contamination. As oil works in a system, sludge and acids form because of a chemical reaction to water, air, heat and pressure. Sludge itself is not normally abrasive but is a source of gummy substances that coat moving parts, clog small openings, and trap abrasive particles suspended in the oil.

Acids, by pitting and corroding, cause roughness in moving parts which in turn causes wear and additional contamination in the oil.

EFFECTS OF CONTAMINATION

All of these contaminants can have a serious effect on the efficiency of a hydraulic system.

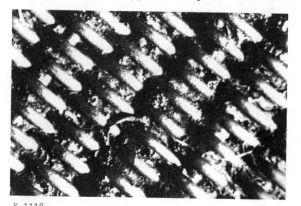

X 1118

Fig. 15—A 100 Micron Filter Completely Plugged

Water, even in small amounts, rusts polished metal surfaces and helps form sludge that coats moving parts and plugs filters (Fig. 15). Plugged filters can cause increased circulation of dirty oil or may even starve the hydraulic components.

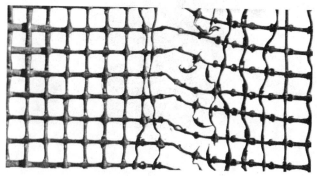

X 1119

Fig. 16—A 150 Micron Filter That Has Been Chemically Corroded

Acids that sometimes form with sludge can corrode parts like the wire mesh filter shown in Fig. 16. This adds still more damaging particles to the oil.

Metallic and non-metallic particles circulating in the oil cause damage that is usually quite apparent. Large particles get caught on the edges of moving

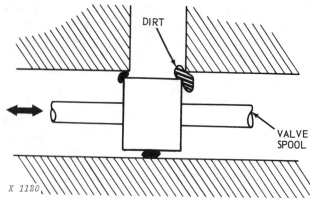

X 1120

Fig. 17—Dirt Will Wear Valve Spools

parts and increase the buildup of sludge or wear off sharp metering edges in valves (Fig. 17). Smaller particles become trapped between closely-fitted parts, causing them to stick or freeze completely. Other particles embed themselves in softer metals to grind away moving parts and seals, causing internal leakage and loss of efficiency.

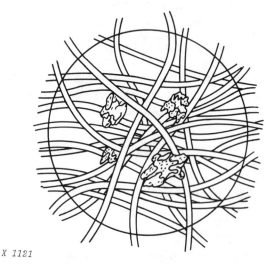

X 1121

Fig. 18—Fibers Can Plug an Orifice

Tiny fibers or lint from wiping cloths or even from your clothes may mat together in tiny pores and orifices (Fig. 18). The fibers themselves do little harm but the solid particles that they trap can plug and wear hydraulic components.

The important thing to remember about contamination is that the damage it causes is not constant. Each dirt particle is an abrasive "seed" that produces more contamination which finally results in permanent damage to the machine.

MAINTENANCE OF FILTERS

Now that we have identified the reasons for contaminated oil and the problems it causes, what are the remedies or solutions to these problems?

As pointed out before, the purpose of the filter in a hydraulic system is to trap and remove dirt from the oil. It is sometimes difficult then to understand why the use of an oil filter will not extend the period between recommended oil changes. Just remember that oil is constantly being contaminated with new dirt from external and internal sources. Because the capacity of a filter is limited, it can only maintain the oil's original quality for a period of time that has been carefully determined by the manufacturer. It cannot extend the life of the oil.

Like a sponge, the capacity of an oil filter is limited. Only particles above a certain size are removed and only one or two of the elements that cause sludge can be filtered. Smaller particles and the rest of the sludge circulate in the system and can only be removed by draining. As filters remain in use, they begin to plug with the particles they are removing and eventually cease to function at all.

Therefore, the only real solution to contamination problems is a regular program of hydraulic system maintenance.

Faithfully follow the manufacturer's recommendations for draining to remove contaminants that the filter cannot remove. This is especially important during the break-in period because the filter will plug much more rapidly than it will during normal operation.

⚠ **CAUTION: Be careful not to burn your hands on hot oil. Let system cool before servicing.**

Be extremely careful about cleanliness when servicing a hydraulic system. No filter can entirely compensate for dirty disassembly or assembly practices. A dirty component or a dirty oil funnel will contaminate the rest of the system and, as we just learned, a small amount of dirt will generate more and more dirt through wear.

Handle filters carefully. A bent or punctured filter will not do the job it is supposed to do. Keep seals in place and store new filters so they are clean when they are installed.

Fig. 19—What Can I Do About Contamination?

Follow the manufacturer's recommendation and clean or replace the filter when it becomes plugged. If the machine does not have a plugged filter indicator, pay strict attention to the filter service intervals shown in the operator's manual. And, finally, use only the filter and oil that is right for that particular hydraulic system.

TEST YOURSELF

QUESTIONS

1. What is the main reason filters are needed in a hydraulic system?

2. Compare full-flow and bypass filtering of hydraulic systems.

3. What does a filter relief valve do?

4. How do surface and depth filters differ?

5. Why are **ad**sorbent filters seldom used in a hydraulic system?

6. What measurement is used to determine the degree of filtration?

7. Explain the term "abrasive seed" as used for dirt in a hydraulic system.

8. What must be done to reduce the possibility of contamination when working with open lines or connectors?

RESERVOIRS, OIL COOLERS, HOSES, PIPES, TUBES, AND COUPLERS / CHAPTER 8

INTRODUCTION

This chapter covers the parts which connect the hydraulic system and store and cool the oil. While less complex than the other parts of the system, these components are vital to the working of the system. Many of them require special services which every serviceman must know.

RESERVOIRS

Every hydraulic system must have a reservoir. The reservoir not only stores the oil, it also helps keep the oil clean, free of air, and relatively cool.

CAPACITY OF RESERVOIRS

A reservoir should be compact, yet large enough to:

• Hold all the oil that can drain back from the system into the reservoir by gravity flow.

• Maintain the oil level *above* the suction line opening.

• Dissipate excess heat during normal operation. (See also "Oil Coolers," which follows in this chapter.)

• Allow air and foreign matter to separate from the oil.

FEATURES OF RESERVOIRS

To serve its purpose, the reservoir must have several features (Fig. 1).

1. The FILLER CAP should be air tight when closed, but may contain the AIR VENT which filters air entering the reservoir to provide a gravity push for proper oil flow. The air vent filter must be kept clean to prevent partial vacuums which restrict gravity flow from the reservoir.

NOTE: Ideally, a system may be designed with a sealed reservoir and no air vent. However, since most systems have changing oil levels and temperatures and different piston sizes, air venting is needed.

2. The OIL LEVEL GAUGE shown in Fig. 1 gives the level of oil in the reservoir without opening it. Dipsticks are still widely used, however.

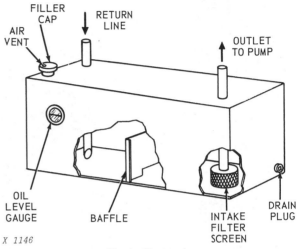

X 1146

Fig. 1 – Reservoir

3. The BAFFLE helps to separate return oil from that entering the pump. This slows the circulation of oil, gives the return oil time to settle, and prevents constant reuse of the same oil. However, no baffle is needed in many modern systems because the same separation of inlet and return oil is achieved by placement of the lines and filters.

4. The OUTLET and RETURN LINES are designed to enter the reservoir at points where air and turbulence are least. They can enter the reservoir at the top or sides, but their ends should be near the bottom of the tank. If the return is above the oil level, the return oil can foam and draw in air.

NOTE: Be careful when placing extra returns from auxiliary equipment in the reservoir. If not placed correctly, they can cause foaming of return oil.

5. The INTAKE FILTER is usually a screen used in series with the system oil filter, which may also be installed in the reservoir. Refer to Chapter 7, "Oil Filters," for details on these components.

6. The DRAIN PLUG allows all oil to be drained from the reservoir. Some drain plugs are magnetic to help remove steel particles from the oil.

LOCATION OF RESERVOIRS

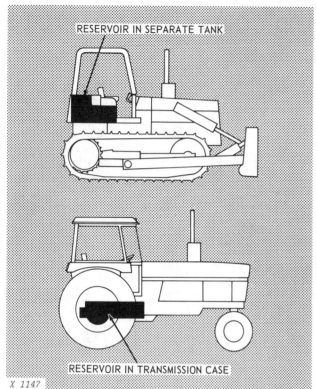

Fig. 2 — Location of Reservoirs on Two Types of Tractors

On modern farm and industrial machines, the reservoir must be compact and light. The bulldozer shown in Fig. 2 has a separate mounted tank while the wheel tractor uses its transmission case as a reservoir. The location of the reservoir depends upon the design of the machine—the available space and the size of the reservoir.

OIL COOLERS

On modern high-pressure systems, cooling the oil can be quite a problem. Often the normal circulation of the oil in the system will no longer do the job. This is why oil coolers are becoming more common on modern equipment.

Two types of oil coolers are widely used:

• **Air-to-Oil Coolers**

• **Water-to-Oil Coolers**

Fig. 3 compares the operation of the two coolers.

AIR-TO-OIL COOLERS use moving air to dissipate heat from the oil. On mobile machines, the cooling system (radiator) fan may supply the air blast (see

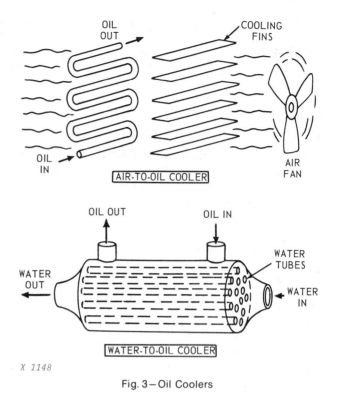

Fig. 3—Oil Coolers

Fig. 4). The cooler has fins which direct the air over long coils of oil tubes which expose more oil to the air. The cooler may also have a tank to store a reserve of cooled oil. A bypass valve is also sometimes used as a safety valve in case the cooler oil tubes become clogged.

WATER-TO-OIL COOLERS use moving water to carry off heat from the oil. The water flows through many tubes and the oil circulates around the cooling tubes as shown. On mobile machines, water from the engine radiator is often used for cooling.

Another less common type of water-to-oil cooler uses the evaporation of water to cool oil. Water is sprayed over coils of oil tubes, while forced air is blown in from the bottom. Part of the water evaporates, cooling the remaining water which in turn draws heat from the oil in the tubes. This cooler is not as compact as the one described above.

LOCATION OF OIL COOLERS

Air-to-oil coolers such as the one shown in Fig. 4 are usually mounted in front of the engine radiator, making use of the fan's air blast. Other coolers can be mounted in a variety of locations, but usually near the reservoir or the engine cooling fan.

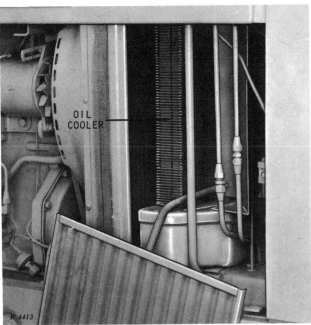

Fig. 4—Location of a Typical Oil Cooler

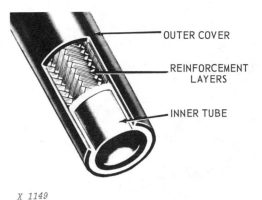

X 1149

Fig. 5—Flexible Hose Construction

FLEXIBLE HOSES

Hoses are the best form of hydraulic plumbing for most uses. Not only do hoses allow for motion, they absorb vibrations and noise, withstand pressure "surges," and are easy to route and connect.

Hydraulic hose is composed of three basic parts (Fig. 5):

- **Inner Tube**
- **Reinforcement Layers**
- **Outer Cover**

The INNER TUBE is a synthetic rubber layer which is oil-resistant. It must be smooth, flexible, and able to resist heat and corrosion.

The REINFORCEMENT LAYERS vary with the type of hose. These layers (or plies) are constructed of natural or synthetic fibers, or braided wire, or a combination of these. The strength of this layer depends upon the pressure requirement of the system where the hose is used.

The OUTER COVER protects the reinforcement layers. A special rubber is most commonly used for the outer cover because it resists abrasion and exposure to weather, oil, and dirt.

Hoses also use metal couplers at each end. These will be covered later in this chapter under "Hose Couplers."

HOW TO SELECT HOSES

To select the proper hose, you should know:

1. The *flow* of the system to find out what *size* of hose is needed.

2. The *pressures* and *temperatures* in the system to determine the *type* of hose to use.

Remember that the size of the hose should match the flow requirements of the system.

Too-small hoses restrict flow, cause overheating and pressure losses.

Too-large hoses may be too weak for the pressure of the system. This is because larger hoses must be stronger to withstand the same pressure as smaller ones. Also, larger hoses are more expensive.

Another factor to consider: the hose must be compatible with the fluid in the system.

Selecting the Type of Hose

Hoses are typed by the strength of their wall construction. The four types are:

- **Fabric Braid**
- **Single Wire Braid**
- **Double Wire Braid**
- **Spiral Wire**

Higher pressure hoses use stronger reinforcing layers or extra layers (Fig. 6).

However, the pressure a hose will handle varies with its size. A larger hose withstands less pressure than a smaller one of the same construction. This is because it has a larger area exposed to the pressure.

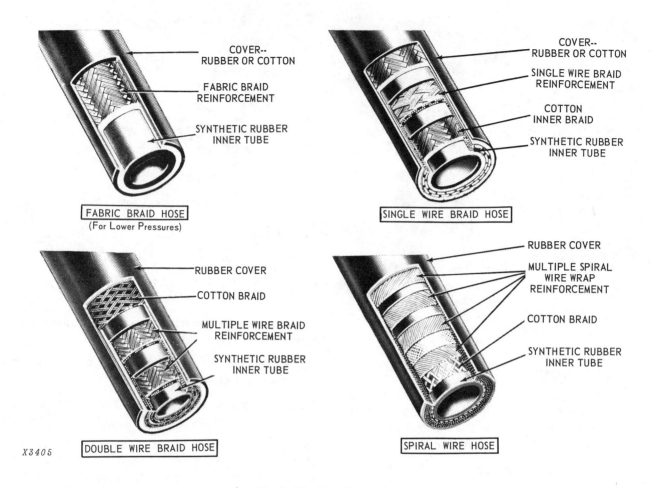

COVER--
RUBBER OR COTTON

FABRIC BRAID
REINFORCEMENT

SYNTHETIC RUBBER
INNER TUBE

FABRIC BRAID HOSE
(For Lower Pressures)

COVER--
RUBBER OR COTTON

SINGLE WIRE BRAID
REINFORCEMENT

COTTON
INNER BRAID

SYNTHETIC RUBBER
INNER TUBE

SINGLE WIRE BRAID HOSE

RUBBER COVER

COTTON BRAID

MULTIPLE WIRE BRAID
REINFORCEMENT

SYNTHETIC RUBBER
INNER TUBE

DOUBLE WIRE BRAID HOSE

RUBBER COVER

MULTIPLE SPIRAL
WIRE WRAP
REINFORCEMENT

COTTON BRAID

SYNTHETIC RUBBER
INNER TUBE

SPIRAL WIRE HOSE

X3405

Fig. 6—The Four Types of Hoses

The hose pressure rating is based on the *working pressure* of the system. This must allow for the maximum "surges" of pressure during the normal operation of the system.

Temperature of the hydraulic oil is also critical in hose selection. All four types will handle the normal heat of hydraulic operation, but special hoses have been designed for extra high temperatures.

The charts which follow give the constructions and applications of the four types of hoses.

FABRIC BRAID HOSE

Construction

Inner Tube: Black synthetic rubber.
Reinforcement: Woven fiber reinforced with spiral wire to prevent collapse.
Cover: Synthetic rubber, oil- and abrasion-resistant.

Uses

Lines for Petroleum-base hydraulic oils, gasoline or fuel oil. In suction lines or in low-pressure return lines.
Temperature Range: -40°F. to +250°F.
Vacuum: 30 In. Hg.

FABRIC BRAID HOSE—Continued

Construction

Inner Tube: Black synthetic rubber, oil-resistant.
Reinforcement: One-fiber braid.
Cover: Black synthetic rubber, oil- and abrasion-resistant.

Uses

Hydraulic oil return lines only, or general-purpose fuel oil, gasoline, water, anti-freeze mixtures, air, and other chemicals.
Temperature Range: -40°F. to +250°F.

Inner Tube: Black synthetic rubber, oil-resistant.
Reinforcement: Two fiber braids.
Cover: Black synthetic rubber, oil- and abrasion-resistant.

Hydraulic oil return lines only, or general purpose fuel oil, gasoline, water, anti-freeze mixtures, air, and other chemicals.
Temperature Range: -40°F. to +250°F.

IMPORTANT: Fabric braid (low pressure) hoses are NOT RECOMMENDED FOR PRESSURE LINE USE IN HYDRAULICS. Therefore, they will not be included in the chart on selecting hoses which follows these charts.

SINGLE WIRE BRAID HOSE

Construction	*Uses*
<u>Inner Tube</u>: Black synthetic rubber. <u>Reinforcement</u>: Two fiber braids. <u>Cover</u>: Synthetic rubber, oil- and abrasion-resistant.	Hydraulic oil lines, fuel oil, anti-freeze solutions, or water lines. <u>Temperature Range</u>: −40° F. to +250°F.
<u>Inner Tube</u>: Black synthetic rubber, oil-resistant. <u>Reinforcement</u>: One braid of high tensile steel wire. <u>Cover</u>: Black synthetic rubber oil- and abrasion-resistant.	Hydraulic oil lines, fuel oil, gasoline or water. <u>Temperature Range</u>: −40°F. to +250° F.

DOUBLE WIRE BRAID HOSE

Construction	*Uses*
<u>Inner Tube</u>: Black synthetic rubber oil-resistant. <u>Reinforcement</u>: Two braids or more of high tensile steel wire. <u>Cover</u>: Black synthetic rubber, oil- and abrasion-resistant.	High-pressure hydraulic oil lines, fuel oil, gasoline or water lines.
<u>Inner Tube</u>: Black, synthetic rubber. <u>Reinforcement</u>: Two braids or more of high tensile steel wire. <u>Cover</u>: Synthetic rubber oil- and abrasion-resistant green color.	Hydraulic lines using phosphate ester base fluids. (Should <u>not</u> be used with <u>petroleum</u> oils.) <u>Temperature Range</u>: −40° F. to +200° F.

NOTE: Both single and double wire braid hoses of the first type are widely used on farm and industrial equipment hydraulic systems.

SPIRAL WIRE HOSE

Construction	*Uses*
<u>Inner Tube</u>: Black synthetic rubber, oil-resistant. <u>Reinforcement</u>: Multiple spiral of high tensile steel wire and one fiber braid. <u>Cover</u>: Black synthetic rubber, oil- and abrasion-resistant.	Very high-pressure hydraulic oil lines or fuel oil lines. <u>Temperature Range</u>: −40° F. to +200° F.

IMPORTANT: Spiral wire hose is recommended where high surge peaks are encountered. Surges can cause weak spots in the wire braids of the less-strong hoses. The spiral wire reinforced hose does not weaken under high surges.

Summary: Selecting Hoses

The following chart will help you to select the proper hose for any pressure application. Find the size of the hose you need and read across to the system working pressure nearest your application. If you find it in column 1, use a single wire braid hose. If in column 2, use a double wire braid hose, or in column 3 use a spiral wire hose.

SELECTING HOSE FOR VARIOUS PRESSURES

Hose Size in Inches	1. Use SINGLE WIRE BRAID Hose if System Working Pressure Equals . . .	2. Use DOUBLE WIRE BRAID Hose if System Working Pressure Equals . . .	3. Use SPIRAL WIRE Hose if System Working Pressure Equals . . .
¼"	3000 psi	5000 psi	— — —
⅜"	2250 psi	4000 psi	5000 psi
½"	2000 psi	3500 psi	4000 psi
⅝"	1750 psi	2750 psi	— — —
¾"	1500 psi	2250 psi	3000 psi
1"	800 psi	1875 psi	3000 psi
1¼"	600 psi	1625 psi	3000 psi
1½"	500 psi	1250 psi	3000 psi
2"	350 psi	1125 psi	2500 psi

Note again how larger hoses are recommended for lower pressures than smaller ones of the same construction.

HOSE FAILURES

When hoses fail prematurely, look for: cracking or splitting, pin hole leaks, improper hose length, rubbing, heat, twisting, wrong hose selection, wrong fittings, or improper routing.

CRACKING OR SPLITTING of the outer cover is common and does not always mean the hose is ruined. The *depth* of the break is what is important. This should be watched for regularly on high-pressure circuits.

PIN HOLE LEAKS are often hard to detect. But a small leak can add up in a few weeks, and can also create fire and skin puncture hazards (see "CAUTION" on page 11—6).

IMPROPER HOSE LENGTH can mean that a too-short hose is stretched under pressure or a too-long one is exposed to hazards from moving parts. Either way the hose is likely to be damaged.

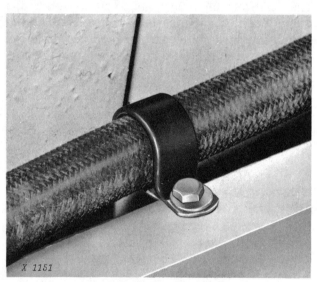

X 1151

Fig. 7—Hose Clamp

RUBBING wears out hose covers, weakens the reinforcement layers, and so wears out the hose prematurely. Clamp or bracket hoses so they cannot rub, or use hose guards.

HEAT from engine exhaust system and radiator can damage hoses. Be sure all hoses are routed away from hot parts or at least are bracketed against contact, and shielded against radiant heat.

TWISTING can restrict the flow of oil and also damage the hose. Bending or flexing is part of hose design; twisting is not. The most common cause of twisting is when one end of the hose is improperly connected to a moving part. To correct this, clamp the hose at the point where it starts twisting, divid-

ing the motion of the hose into two planes. If some twisting cannot be avoided, allow as much free hose as possible. When connecting a hose, hold it to avoid twisting as you tighten the end fittings.

WRONG HOSE SELECTION happens when hoses are used that are of the wrong size or pressure rating. See the chart on selecting hoses in this section. Remember, *Don't Economize!* Selecting the wrong hose can cause many of the failures above.

WRONG FITTINGS occur when replacement fittings are not matched to the hoses in size or style. For details, see "Hose Couplers" later in this chapter.

COLLAPSED SUCTION HOSE is another type of breakdown, seemingly simple. But it is possible that the inner layer of rubber of the suction hose will, when it starts to deteriorate, collapse inward, completely sealing off flow, without showing any outer symptoms of collapse. A noisy pump, lack of pressure, "spongy" action, or no action at all are indications of a collapsed suction tube.

IMPROPER ROUTING is the No. 1 cause of hose failure. Included here are twisting and abrasion, but also sharp bends, hoses which are too long or short, use of too many fittings, wrong-way connections. Since there are so many possible causes, care must be taken to find the right remedy to this problem. To prevent improper routing, study the next part of this chapter, "Installing Hoses."

INSTALLING HOSES

Now that we have discussed hose failures, let's talk about the leading remedy—proper hose installation.

Follow the six basic rules when installing a hose:

1. **Avoid Taut Hose.** Even where the hose ends do not move in relation to each other, allow some slack to prevent strain. Taut hoses tend to bulge and weaken under pressure.

2. **Avoid Loops.** Use angled fittings to prevent long loops. Doing this cuts down the length of hose needed and makes a neater installation.

3. **Avoid Twisting.** Hoses are weakened and fittings are loosened by twisting either during installation or machine operation. Use a hose clamp or allow some free hose where necessary. And remember—tighten the fitting on the hose—not the hose on the fitting.

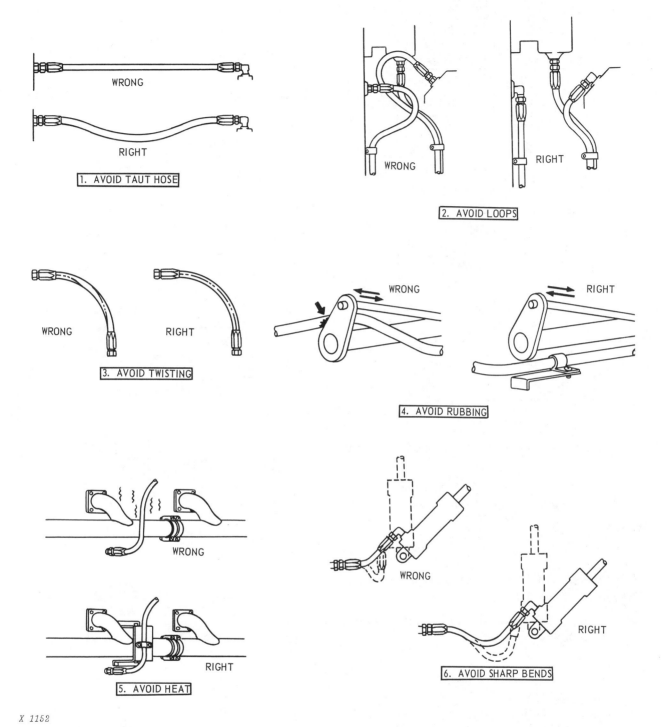

X 1152

Fig. 8—Installing Hoses

4. Avoid Rubbing. Clamp or bracket hoses away from moving parts or sharp edges. If this doesn't work, use a hose guard of wire spring or flat armor spring design.

5. Avoid Heat. Keep hoses away from hot surfaces such as engine manifolds. If you can't route the hoses away from these areas, shield them.

6. Avoid Sharp Bends. The bend radius depends upon the hose construction, size, and pressure. The manufacturer recommends a certain limit for bending on each hose. At lower pressures, a tighter bend is permitted. Where possible, reroute hoses to avoid sharp bends. Or, allow extra slack but watch for kinks or loops. *Remember: Only the hose is flexible, not the fittings.*

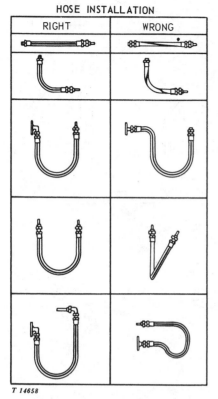

Fig. 9—How to Route Hoses

For a quick check on how to route hoses, refer to Fig. 9.

In summary, route hoses for good appearance. Good appearance and good function go together.

HOSE COUPLERS

Hose couplers include two types:

- **Fittings—part of the hose, they have a socket and nipple or sleeve.**

- **Adapters—a separate part for joining the hose fitting to another line.**

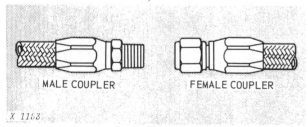

Fig. 10—Male and Female Couplers

Fittings and adapters are called either male or female couplers. The hollow female coupler mates with the male type (Fig. 10).

Hose couplers are made of steel, brass, stainless steel or, in a few applications, plastic. Steel is generally used because it withstands both high pressures and heat.

Let's discuss each type of hose coupler.

Hose Fittings

Hose fittings can seal in many ways. The five major methods are shown in Fig. 11.

Besides straight fittings, elbow fittings are also available.

Elbows should be used for access to hard-to-reach connections and for special routing problems.

1. SAE Straight Thrd. O'Ring

2. Male Pipe

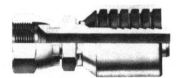

3. 37°JIC

4. Code 61/62 SAE Flanges

Fig. 11—Five Major Ways of Sealing Hose Fittings

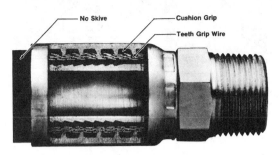

Fig. 12—Hose Fittings

Hose fittings are also designed to be either *permanent* or *reusable* (Fig. 12, top).

PERMANENT HOSE FITTINGS are discarded with the hose. They are either crimped or swedged onto the hose. Some dealers have crimping machines which can make up hose assemblies using permanent fittings and stock hose cut to length (see Fig. 13).

REUSABLE HOSE FITTINGS are either pushed on, screwed on, or clamped onto the hose. When the hose wears out, the fittings can be removed and used on a new hose cut from stock. Most reusable fittings can be converted to another thread type by changing the nipple in the socket. As might be expected, reusable fittings are slightly more expensive.

Medium and high pressure hose fittings may be the same in appearance. Yet, each must be distinguished for proper use. Always check the fitting manufacturer specifications. Never use a fitting that does not meet the requirements of the job.

If hose and fittings are matched *incorrectly*, then the results can be pinhole leaks, ruptures, heat build ups, pressure drops, cavitation, and other failures.

Installing Permanent Hose Fittings

Permanent hose fittings are installed using a hose crimping machine (Fig. 13).

The crimping machine can be powered by a hand pump, an air pump, or a hydraulic pump.

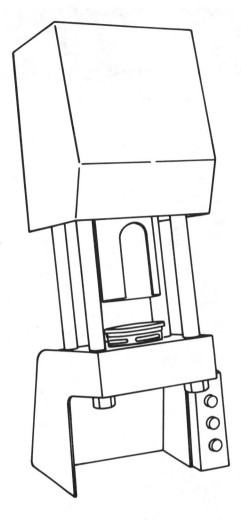

Fig. 13 — A Hose Crimper

The hose is cut, assembled to its fittings, inserted in the machine, and crimped.

The actual hose crimping is done by die fingers powered by a cylinder ram.

The die fingers are adjusted for the size of the hose fitting, while a depth stop is adjusted to suit the length of the fitting. Be sure to follow the specific instructions given with the manual for the crimping machine. We have shown only one typical unit.

 CAUTION: Always follow fitting manufacturer's installation instructions.

SOCKETLESS (Low Pressure)	SOCKET AND NIPPLE (Medium Pressure, Single Wire Braid Hose)	SOCKET AND NIPPLE (High Pressure — Notch on Socket) (Multiple Wire Braid Hose)
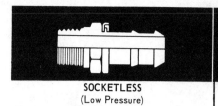	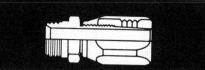	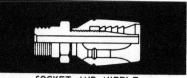

SOCKETLESS (Low Pressure)

ASSEMBLY

1. Cut hose to length with a sharp knife. Oil inside of hose and nipple liberally.

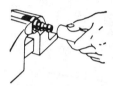

2. Push hose on fitting until it bottoms underneath protective cap.

DISASSEMBLY

1. Slit hose from protective cap to end of nipple.

2. Bend hose and snap it off with a quick pull.

SOCKET AND NIPPLE (Medium Pressure, Single Wire Braid Hose)

1. Cut hose to length with fine tooth hacksaw or cut-off wheel. Skive if required (see at right). Screw hose counterclockwise into socket until it bottoms. Back off ¼-turn.

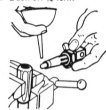

2. For male ends, if required: insert assembly tool mandrel (of correct size) into nipple and oil nipple threads, assembly tool, and inside of hose.

3. For male ends, screw nipple clockwise into socket until snug against socket.

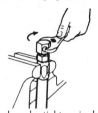

4. For female ends, tighten nipple and nut on assembly tool if required. Screw nipple clockwise into socket. Leave 1/32 in. to 1/16 in. clearance between nut and socket.

SOCKET AND NIPPLE (High Pressure — Notch on Socket) (Multiple Wire Braid Hose)

1. Cut hose to length. To strip off protective cover (if necessary) cut around hose down to metal wire reinforcement, then cut cover lengthwise and pull off cover. Clean wire with wire brush or soft wire wheel. Avoid flaring or fraying wire reinforcement.

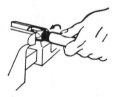

2. Screw hose counterclockwise into socket until it bottoms.

3. Oil nipple threads and inside of hose liberally. Use grease for larger sizes.

4. Screw nipple clockwise into socket. Leave 1/32 in. to 1/16 in. clearance. X 1156

Fig. 14—Installing Reusable Hose Fittings

Installing Reusable Hose Fittings

Reusable hose fittings must be properly installed (Fig. 14). The socketless type for low pressures is shown in column 1. The socket and nipple types are shown in column 2 (medium pressure) and column 3 (high pressure).

Be sure to securely fasten all hose fittings. *The blowoff of a hose at high pressures can be dangerous!*

⚠ **CAUTION: Proper tools and skill are necessary to install reusable hose fittings. Always follow fitting manufacturer's instructions when installing specific fittings.**

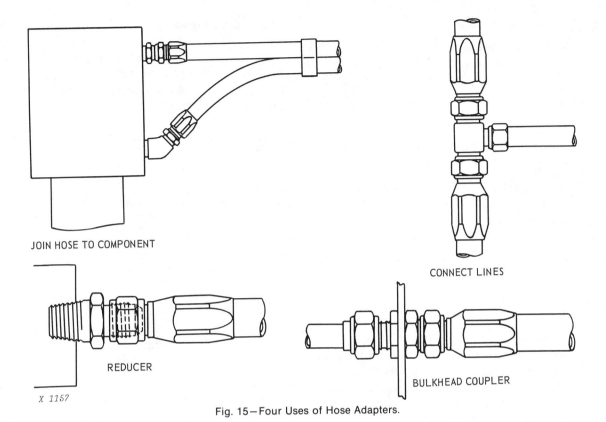

JOIN HOSE TO COMPONENT

CONNECT LINES

REDUCER

BULKHEAD COUPLER

X 1157

Fig. 15—Four Uses of Hose Adapters.

Hose Adapters

A hose adapter is a separate part for joining a hose fitting to another line or fitting.

Since most hydraulic components have ports with tapered pipe threads, adapters are often needed for proper mating.

Adapters are used in four ways (Fig. 15):

• To join a hose fitting to a component.

• To connect two or more lines or fittings.

• To replace a bushing, as a reducer.

• To both anchor and connect lines (bulkhead coupler).

Installation of Hose Couplers

Besides the rules for installing hoses we covered earlier, there are twelve rules for good hose coupler installations:

1. Be sure working pressure rating of coupler corresponds to working pressure rating of hose.

2. Be sure the seal replacement is matched to the mating coupler.

3. Use flared adapters or elbow hose fittings where possible instead of pipe adapters.

4. Improve line routing by use of 45° and 90° adapters or elbows.

5. Attach male ends of hose assemblies before female ends.

6. Tighten swivel nuts only until snug—do not overtighten.

7. Tighten only nipple hex nut and not the socket.

8. Use pipe sealing compound on male threads only—make sure compound is compatible with the hydraulic oil.

9. Use open-end wrenches for assembly—do not use pipe wrenches.

10. Use two wrenches where necessary to prevent twisting of hoses.

11. Tighten the fitting on the hose—not the hose on the fittings.

12. As a general rule, tighten the fitting until finger tight, then use wrenches to tighten the fitting two extra flats. If leakage occurs after operation, tighten one extra flat.

Failures of Hose Couplers

Leakage is the most common failure of hose couplers. Leakage is usually the result of stripped threads, damaged O-rings, or mis-matched seals.

In some cases, failures have resulted from improper assembly of fittings in the hose ends.

Premature failure also can result from overtightening of swivel nuts while leaving pipe threads loose.

Other possible failures result in using too much sealing compound ("pipe dope") which restricts flow and contaminates the hydraulic oil.

Cracked coupler sockets may result from using a low-pressure socket in a high-pressure system.

PIPES AND TUBES

The choice between pipe or tubing depends on system pressure and flow. The advantages of tubing include easier bending and flaring, fewer fittings, better appearance, better reusability, and less leakage. However, pipe is cheaper and will handle larger volumes under higher pressures. Pipe is also used where straight-line hook-ups are needed, and for more permanent installations.

In either case, the hydraulic lines must be compatible with the entire system. Pressure loss in the lines must be kept as small as possible for an efficient circuit.

CONSTRUCTION OF PIPES AND TUBING

Pipes for hydraulic lines should be made of seamless cold-drawn mild steel. *Galvanized pipe should NOT be used because the zinc coating may flake or scale and damage the valves and pumps.*

Tubing can be made from a variety of materials:

Copper—The use of copper is limited to low-pressure hydraulic systems where vibration is limited. Also, copper tends to become brittle when flared and subjected to high heat.

Aluminum—This tubing is also limited to low-pressure use, yet has good flaring and bend characteristics.

Plastic—Plastic tubing lines are made from a variety of materials; nylon is the most suitable. For use in low pressure hydraulic application only.

Steel—Tubing constructed of cold-drawn steel has become the accepted standard in hydraulics

where high pressures are encountered. There are two types of steel tubing; seamless and electric welded. Seamless is produced by a cold drawing of pierced or hot extruded billets. Welded is made by forming a cold-rolled piece of steel into a tube, then welding and drawing it.

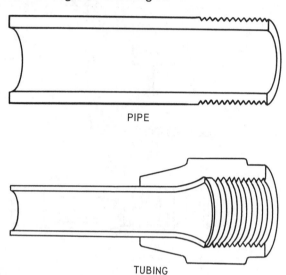

PIPE

TUBING

X 1158

Fig. 16—Pipes and Tubing Compared

SELECTION OF PIPES AND TUBING

The thickness of the wall determines the strength of a pipe or tube. The thicker the wall, the stronger the line.

However, when replacing pipes or tubes, consider both the system pressure and the size of the line.

Remember this when selecting pipes or tubes:

Too-small lines may cause pressure drops, restrict flow, and create heat—a power loss.

Too-large lines are cumbersome and costly.

Be sure that the pipes or tubes match the hoses used in the same pressure circuit. The *inside* diameter of the lines is the critical factor in matching them for size.

FAILURES OF PIPES AND TUBING

If rigid lines are of good quality and are well maintained, failures should rarely occur.

Watch for loose clamps, which cause vibrations.

Check lines that may have been accidently hit, bent, or pinched during operation.

Check for "wet spots" that could mean a pin-hole leak in the line.

INSTALLING PIPES

1. Always replace pipes with ones of identical design and material.

2. Where possible, *avoid straight-line hookups,* especially in short runs. The reason is that straight runs do not allow for enough expansion or contraction during heat and pressure changes.

3. When installing long pipe lines, use brackets and clamps to reduce strain and fatigue. All components or heavy fittings should also be bolted down to eliminate stress on the pipe.

4. For convenience of routing utilize bulkhead fittings where a line passes through a wall or beam. This will not only provide ease of dismantling, but also gives added support.

5. Replacement pipe must be clean and free from rust and scale. To achieve a bright and clean inside surface, two methods are employed by the manufacturer: sand blasting and pickling.

INSTALLING TUBING

1. Always replace tubing with parts of the same material and design. *Do not substitute.*

2. Plan to use as few fittings as possible by using bends in the tubing where practical.

3. Plan the general path of the tubing as follows:

   a. Use the simplest bends, the fewest bends, and the least sharp bends possible.

   b. Avoid points of interference with the operator, the moving parts, and any access doors or controls. Keep the lines from protruding if at all possible.

   c. *Avoid straight-line hookups where possible.* These runs make the line difficult to remove and do not allow for enough expansion and contraction.

d. Support long tubing runs with brackets or clamps. Group lines and clamp several for neater appearance.

e. Do not route tubes through bulkheads or walls if you can avoid it. If you must, use bulkhead connectors for easier removal and added support.

4. Install the tubing as follows:

   a. Install the couplers on the components to be connected.

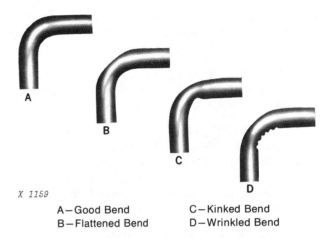

X 1159

A—Good Bend C—Kinked Bend
B—Flattened Bend D—Wrinkled Bend

Fig. 17—Good and Bad Bends in Tubing

   b. Line up the tubes and decide exactly where the bends should be made. Use the proper tube bending tools to prevent flattened, kinked, or wrinkled bends (Fig. 17). The bends should be accurate and smooth so they do not restrict flow. As a general rule, the radius of the bend should be three to five times the diameter of the tubing, but follow the manufacturer's recommendations.

   c. Refer to Fig. 18 for a guide in routing tubes properly.

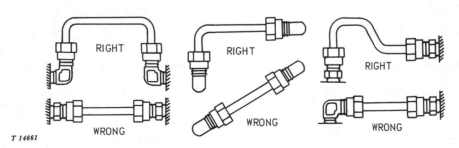

T 14661

Fig. 18—Routing of Tubes

PIPE COUPLERS

Pipe is normally connected by threading its outside diameter and screwing it into a tapped hole. Two types of threads have become accepted standards for hydraulic applications: The American or National Standard Tapered Thread (NPT) and Dryseal American or National Taper Pipe Thread (NPTF). Fuel/Fnc

In NPT threads, interference contact between thread flank does the sealing. These threads require pipe sealing compound to eliminate leaks.

In NPTF threads, the roots and crests engage before the flanks touch. When tightened, the crests are crushed, creating the seal.

Although NPT and NPTF threads are interchangeable, both male and female dryseal threads must be mated to obtain effective sealing.

Besides using pipe sealing compounds, pipe threads and connections can also be sealed by using fittings with pipe lock nuts. These fittings have a free nut with a Teflon seal ring. When positioned and tightened, the nut forms the seal and locks the fitting into position.

Pipe fittings are available in many different styles, both male and female. They are constructed of three types of material:

1. Brass in small sizes for low and medium pressures.

2. Cast iron in large sizes for low and medium pressures.

3. Steel, of varying sizes for hydraulic applications where high pressures are encountered.

Installation of Pipe Couplers

Here are some helpful hints to assist you in proper installation of pipe couplers:

1. When cutting pipe threads, use sharp standard pipe taps or dies, and a good cutting oil. This keeps leakage to a minimum.

2. Remove burrs from both the inside and outside of the pipe end.

3. Clean all foreign matter from pipes and fittings.

4. Use union connectors at various points to aid in later removals.

5. When using pipe sealing compound cover only two-thirds of the threaded male end. *Never apply pipe compound to female threads.* Make certain the pipe compound is compatible with the hydraulic fluid, and *never use shellac as a sealer.*

TUBE FITTINGS

Most tube fittings are joined to the tube by a union which allows the fitting to be tightened while the tube remains fixed. (A few tubes have fittings welded to the tubing.)

Many types of fittings are available, but the main difference is in the way they seal. The two basic types of seals are *flared* and *flareless* (Fig. 19).

Flared Fittings

Flared fittings are used with thin-walled tubing that is easily flared. Sealing is by metal-to-metal contact. The flared end of the tube is squeezed between the mating surfaces as the fitting is tightened firmly to another connection. The flare angle is either 37° or 45°. The 37° flare is the accepted standard for farm and industrial hydraulics, and is available only in steel. The 45° flare is widely used by the auto industry for low pressure circuits and is normally made of brass.

Flared Fittings come in several types (Fig. 19).

1. The standard THREE-PIECE FLARED fitting has a body, a sleeve, and a nut which fits over the tube. The free-floating sleeve allows clearance between the nut and the tube, aligns the fitting, and is a lock washer for the tightened assembly. Advantages are the locking action of the sleeve, plus the fact that the flared tube is not rotated and so worn during assembly.

2. TWO-PIECE FLARED fittings have no sleeve but use a tapered nut to align and seal the flared end of the tube. This fitting has some disadvantages: when tightened, it tends to bind on the flare which may cause unequal sealing. This same friction may twist the tubing somewhat.

3. INVERTED-FLARE fittings have a 45° flare on the inside of the fitting body. This type is used primarily in the auto industries.

4. SELF-FLARING fittings are designed with a wedge-type sleeve. When the nut is tightened, the wedge presses against the tube end and a female part of the fitting to form the flare. This fitting is strong and resists vibration with a minimum of tightening.

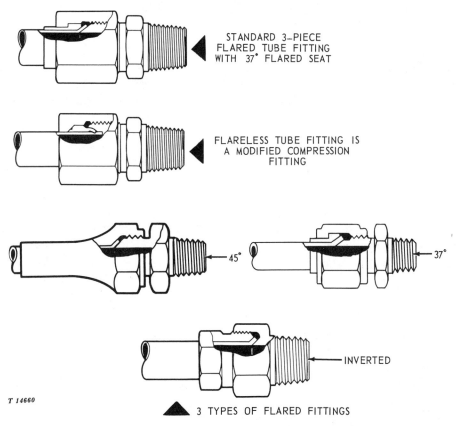

STANDARD 3-PIECE
FLARED TUBE FITTING
WITH 37° FLARED SEAT

FLARELESS TUBE FITTING IS
A MODIFIED COMPRESSION
FITTING

←—45°

←—37°

←— INVERTED

T 14660

3 TYPES OF FLARED FITTINGS

Fig. 19—Tube Fittings

Flareless Fittings

The advantage of flareless fitting is that they require no special tools to flare the tubing. They are not limited by tube size; and they are reusable. There are three basic types of flareless fittings:

1. FERRULE FITTINGS are most commonly used. They are composed of three components—body, nut and ferrule. The wedge-shaped ferrule is drawn down by tightening the nut, creating a positive seal between the body and the ferrule. At the same time, the cutting edge of the ferrule cuts into the tube wall, creating another positive seal.

2. INVERTED FLARELESS fittings use the ferrule sealing method, but the seal only takes place in part of the boss. A male thread nut is used to tighten the ferrule into position. The advantage to this fitting is that it reduces by one the number of potential leak points.

3. COMPRESSION fittings are limited to use with thin-walled tubing. They crimp the end of the tube to seal. However, vibrations can work the nut loose in some applications.

Another type of compression fitting seals at both ends of the sleeve, crimping the tube twice between the fitting body and the nut. This type of fitting is restricted. Use with thin-walled tubing where there are no vibrations and pressure is low.

4. O-RING fittings have the advantage of a replaceable sealing element. Also, the condition of the tube ends is not critical since they do not seal.

Tightening Tube Fittings

The most important rule for tightening tube fittings is: *Tighten only until snug. Do not overtighten.*

Where necessary, use two wrenches on the fittings to avoid twisting the lines.

If a fitting starts to leak and appears loose, *retighten only until the leak stops.*

More damage has been done to tube fittings by overtightening than from any other cause.

The chart on the next page shows how to tighten flare-type tube fittings.

HOW TO TIGHTEN FLARE-TYPE TUBE FITTINGS

Line Size (Outside Diameter)	Flare Nut Size (Across Flats)	Tightness (Ft-lbs)	Recommended Turns of Tightness (After Finger Tightening)	
			Original Assembly	Re-assembly
3/16"	7/16"	10	1/3 Turn	1/6 Turn
1/4"	9/16"	10	1/4 Turn	1/12 Turn
5/16"	5/8"	10-15	1/4 Turn	1/6 Turn
3/8"	11/16"	20	1/4 Turn	1/6 Turn
1/2"	7/8"	30-40	1/6 to 1/4 Turn	1/12 Turn
5/8"	1"	80-110	1/4 Turn	1/6 Turn
3/4"	1 1/4"	100-120	1/4 Turn	1/6 Turn

FOUR BOLT FLANGE FITTINGS

Flange fittings are widely used in industrial machine application, mainly because of the large diameter oil pipes required. Another advantage is the relative ease with which they can be installed.

There are actually two flanges involved.

- **A flange that is welded or brazed to the oil pipe. This flange has an O-ring groove cut into its face.**

- **A flange that fits over the pipe flange and holds it against a flat mating surface by means of four cap screws, and compresses the O-ring seal or packing to complete the sealing. This flange may be a one-piece or two-piece (split) design.**

Servicing Four Bolt Flange Fittings

Procedure is shown in Fig. 20.

1. Clean sealing surfaces (A). Inspect. Scratches cause leaks. Roughness causes seal wear. Out-of-flat causes seal extrusion. If the defects cannot be polished out, replace the component.

2. Install the O-ring (and backup washer if required) into the groove using petroleum jelly to hold it in place.

3. Split flange: Loosely assemble split flange (B) halves. Make sure the split is centrally located and perpendicular to the port. Hand tighten the cap screws to hold parts in place. Do not pinch O-ring (C).

4. Single piece flange (D): Place the hydraulic line in the center of the flange and install the cap screws. The flange must be centrally located on the port. Hand tighten the cap screws to hold the flange in place. Do not pinch the O-ring.

5. Tighten one cap screw, then tighten the diagonally opposite cap screw. Tighten the two remaining cap screws.

DO NOT use air wrenches. DO NOT Tighten one cap screw fully before tightening the others. DO NOT over tighten.

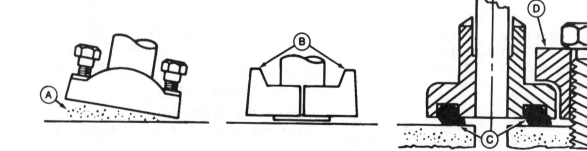

A—Sealing Surface   B—Split Flange   C—Pinched O-Ring   D—Single Piece Flange

Fig. 20—Four Bolt Flange Fitting Service Recommendations

QUICK DISCONNECT COUPLERS

Quick disconnect couplers are used where oil lines must be connected or disconnected frequently. They are self-sealing devices and do the work of two shutoff valves and a tube coupler.

These couplers are fast and easy to use and keep oil loss at a minimum. More importantly, there is no need to drain or bleed the system every time a hookup is made. However, dust plugs must be inserted in the coupler ports when the oil lines are disconnected.

Quick couplers consist of two halves: the body contains a spring-loaded poppet or seal, while the other half is inserted to open the poppet when the two halves are connected. A locking device holds the two halves and seals them.

There are four basic types of quick couplers:

- **Double Poppet**
- **Sleeve and Poppet**
- **Sliding Seal**
- **Double Rotating Ball**

DOUBLE POPPET couplers (Fig. 21) have a self-sealing poppet in each coupler half. When closed, the poppets seal in oil. When connected, the poppets push each other off their seats to allow oil flow. When disconnected, the poppets close again by spring action before the two halves release their seal. The coupler halves are locked in place by a ring of balls which are held into a ring in the inserted coupler half by a spring-loaded outer sleeve.

SLEEVE AND POPPET couplers have a self-sealing poppet in one half and a tubular valve and sleeve in the other. The sleeve is inserted first and gives an added margin of sealing against oil loss or air entry.

SLIDING SEAL couplers have a sliding gate which covers the port in each coupler half while disconnected. This type of coupler spills more oil during hookup than the others.

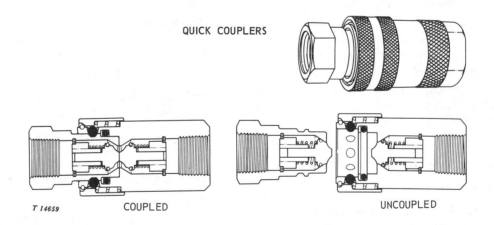

QUICK COUPLERS

T 14659   COUPLED   UNCOUPLED

Fig. 21 — Quick Disconnect Couplers

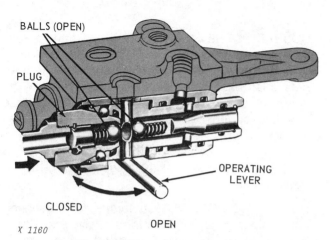

BALLS (OPEN)

PLUG

OPERATING
LEVER

CLOSED

OPEN

X 1160

Fig. 22 – Quick Disconnect Coupler of Double
Rotating Ball Type

DOUBLE ROTATING BALL couplers (Fig. 22) are connected by inserting the line plug into the coupler body while rotating the lever to the position shown. The lever opens the valve balls in both the plug and body, allowing oil to flow. When the coupler is disconnected, pulling the line plug rotates the lever to close the valve balls without loss of oil.

The locking device for this coupler is the same as for the "double poppet" type. The halves are locked by a ring of small balls held into a groove on the inserted plug by an outer sleeve.

However, the coupler in Fig. 22 has an automatic lock release in case the lines are pulled loose. (This is useful when pulling implements such as a plow behind a tractor, for example. If the plow hits a stump or rock, the hitch releases. The disconnect coupler also releases the hose lines at the same moment without damage.)

NOTE: Couplers are available in two versions:

- **Quick disconnect**

- **Automatic release — which releases the coupler when the hydraulic line is pulled, by force.**

USE CORRECT HOSE TIPS

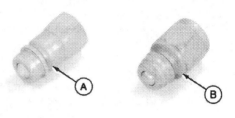

A—ISO Hose Tip   B—Old Hose Tip
Fig. 23 — Different Hose Tips

It is important that the hydraulic line connectors are compatible.

Newer machines have connectors that conform to ISO* and SAE** standards. The hose tips on older equipment may be of an older and different design. If so, the tips must be changed to the new type so the equipment can be hooked up.

 *International Standards Organization
**Society of Automotive Engineers

TEST YOURSELF

QUESTIONS

1. The reservoir takes care of what vital functions besides storing oil?

2. What two mediums are used for cooling in most oil coolers?

3. What are the three basic parts of a flexible hose?

4. Name the six rules of things to "avoid" when installing hydraulic hoses.

5. (True or False) "Galvanized pipe is recommended for high-pressure hydraulic plumbing."

6. (True or False) "The best route for tubing between two points is not a straight line."

HYDRAULIC SEALS / CHAPTER 9

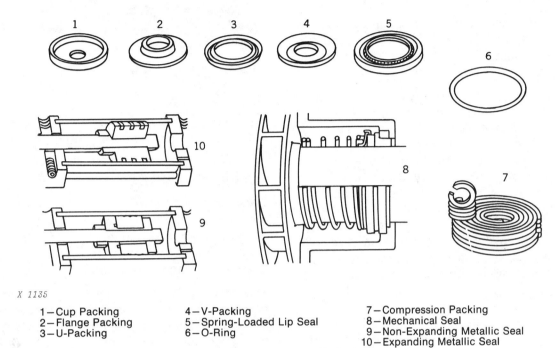

X 1135

1—Cup Packing
2—Flange Packing
3—U-Packing

4—V-Packing
5—Spring-Loaded Lip Seal
6—O-Ring

7—Compression Packing
8—Mechanical Seal
9—Non-Expanding Metallic Seal
10—Expanding Metallic Seal

Fig. 1—Types of Hydraulic Seals

INTRODUCTION

No hydraulic circuit can operate without the proper seals to hold the fluid under pressure in the system. Seals also keep dirt and grime out of the system.

Hydraulic seals appear to be simple objects when held in the hand. But in use they are complex, precision parts and must be treated carefully if they are to do their job properly.

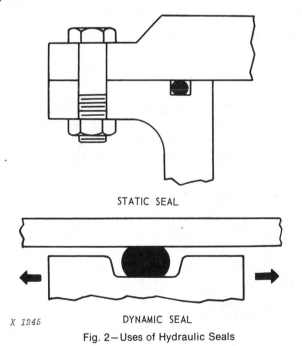

STATIC SEAL

THE USES OF SEALS

Hydraulic seals are used in two main applications:

- **Static Seals—to seal fixed parts**
- **Dynamic Seals—to seal moving parts**

Static seals usually are gaskets, but may also be O-rings or packings (Fig. 2).

Dynamic seals include shaft and rod seals and compression packings. A slight leakage in these seals is permissible for seal lubrication.

X 1245

DYNAMIC SEAL

Fig. 2—Uses of Hydraulic Seals

Later in this chapter we'll talk in more detail about the uses of seals and the problems with each one.

THE TYPES OF SEALS

Seals can be typed by their form or design (Fig. 1). Let's discuss each type of seal.

O-RINGS

The simple O-ring is the most popular seal in farm and industrial hydraulics. Usually made of synthetic rubber, it is used in both static and dynamic applications.

O-rings are designed for use in grooves where they are compressed (about ten percent) between two surfaces. In dynamic use, they must have a smooth surface to work against. O-rings are not used where they must cross openings or pass corners under pressure. Nor are they used on rotating shafts because of wear problems. In static use, under high pressure, they are often strengthened by a back-up ring to prevent them from squeezing out of their grooves. The back-up ring is usually of fiber, leather, synthetic plastic, or rubber design. Leather or fiber should not be used in a cylinder.

U- AND V-PACKINGS

U- and V-packings are dynamic seals for pistons and rod ends of cylinders, and for pump shafts. They are made of leather, synthetic and natural rubber, plastics, and other material.

These packings are installed with the open side, or lip, toward the system pressure so that the pressure will push the lip against the mating surface to form a tight seal.

U- and V-packings are made up of several U- or V-shaped elements and are used in packing glands or cases which hold them in one piece. They are popular for sealing rotating shafts, pistons, and rod ends of cylinders.

SPRING-LOADED LIP SEALS

These seals are a refinement of the simple U- or V-packing. The rubber lip is ringed by a spring which gives the sealing lips tension against the mating surface. Usually the seal has a metal case which is pressed into a housing bore and remains fixed. This seal is often used to seal rotary shafts. The lip normally faces in toward the system oil. Double-lip seals are sometimes used to seal in fluids on both sides of an area.

CUP AND FLANGE PACKINGS

Cup and flange packings are dynamic seals and are made of leather, synthetic rubber, plastics, and other material. The surfaces are sealed by

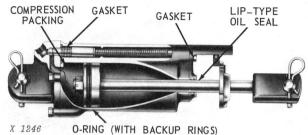

Fig. 3—Uses of Seals in a Hydraulic Cylinder

expansion of the lip or beveled edge of the packing. They are used to seal cylinder pistons and piston rods.

MECHANICAL SEALS

These seals are designed to eliminate some of the problems in using chevron packings for rotating shafts. They are dynamic seals, usually made of metal and rubber. Sometimes the rotating portion of the seal is made of carbon, backed up with steel.

The seal has a fixed outer part attached to the housing. An inner part is attached to the revolving shaft and a spring holds the two parts of the seal tightly together.

A rubber ring (flange-shaped) or a diaphragm is usually included to permit lateral flexibility, and to keep the rotating part of the seal in motion.

METALLIC SEALS

Metallic seals used on pistons and piston rods are very similar to the piston rings used in engines. They may be either expanding or non-expanding. Used as dynamic seals, they are usually made of steel.

Unless fitted very closely, non-expanding seals will leak excessively. Expanding seals (for use on pistons) and contracting seals (for use on piston rods) are subject to moderate friction and leakage losses.

Precision metallic seals, however, are not as subject to leakage, and are especially well-adapted for use in extremely high temperatures.

Since metallic seals are more subject to leakage than others, fluid wiper seals with external drains are often used.

COMPRESSION PACKINGS

Compression packings (jam packings) are used in dynamic applications. They are made of plastics, asbestos cloth, rubber-laminated cotton, or flexible metals.

Compression packings are often used in the same ways as U- and V-packings. They are designed as single coils or as endless rings from which sized pieces may be cut.

Compression packings are generally suitable only for low pressure uses. Lubrication is very important, as they will score moving parts if allowed to run dry.

COMPRESSION GASKETS

Gaskets are suitable, of course, only for static uses. Gaskets seal by molding into the imperfections of the mating surfaces. This molding depends upon a very tight seal at all points.

Gaskets are made of many materials, both metallic and non-metallic, and come in hundreds of shapes.

HOW SEALS ARE SELECTED FOR EACH USE

The designer of a hydraulic system has many factors to consider when choosing a seal. Some of these are:

1. *Will the seal resist all pressures expected?*

2. *Can the seal withstand the heat of operation?*

3. *Will the seal wear out too rapidly?*

4. *Will the seal be harmed by the fluid in use?*

5. *Does the seal fit without dragging on the moving parts?*

6. *Will the seal score or scratch polished metal parts?*

Each application for a seal presents a different set of problems. This is why so many seals are on the market today. And why the right choice of a seal for replacement is so crucial. Always follow the manufacturer's recommendation listed in his parts catalog.

SEAL FAILURES AND REMEDIES

A top-notch hydraulic system which is very complex still depends upon the simple seal for good operation.

The perfect seal should prevent all leakage. But this is not always practical. In dynamic uses, for example, slight leakage as an oil film aids in lubrication of the moving parts. In practice, a seal is regarded as free of leaks if, after continued operation, any leakage is very hard to detect. In other words, there are no drips or pools of oil.

Of course, internal leakage is always hard to detect and this takes some testing to find out if and where the system is leaking.

To get the best use of seals, proper handling and replacement is vital. Most seals are fragile and can be easily damaged.

To prevent this, keep seals protected in their containers until ready for use. Store them in a cool, dry place free of dirt. Seals should be given the same care as precision bearings.

As a general rule, **replace all seals that are disturbed during repair of the system.** The price of a few seals is very cheap compared to a return repair job to correct leaks.

O-RING MAINTENANCE

O-rings can be easily damaged by cutting or nicks from sharp objects. They can also be damaged by heat, improper fluids, inadequate lubrication, and improper installation (Fig. 4).

Installation of O-Rings

1. Be sure the new O-ring is compatible with the hydraulic fluid. Otherwise, the O-ring may corrode, crack, or swell in operation.

2. Clean the entire area of all dirt and grit before installing O-rings.

3. Inspect O-ring grooves before installing rings. Remove any sharp edges, nicks, or burrs with a fine abrasive stone. Then reclean the area to remove all metal particles.

4. Inspect the shaft or spool (if used). Sharp edges or splines can cut O-rings. Remove any nicks or burrs with a *fine* abrasive stone. Then polish with a *fine* abrasive cloth. Reclean the area to remove all metal particles.

5. Lubricate the O-ring before installing it. Use the same fluid as used in the system. Also moisten the groove and shaft using the hydraulic fluid.

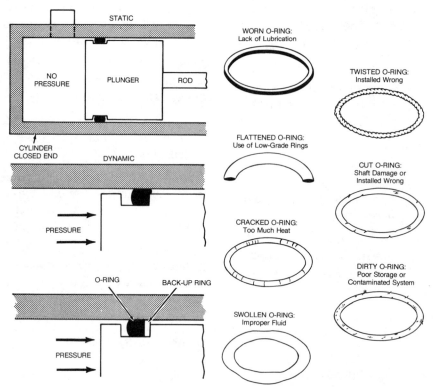

Fig. 4 — O-rings and Teflon Rings Often Replace Flat and Other Type Seals in Cylinders

6. Install the O-ring, protecting it from sharp edges or openings. Be careful not to stretch it more than necessary.

7. Align the parts accurately before mating to avoid twisting or damage to the ring.

8. Check to see that the O-ring is of the correct size to give only a slight "squeeze" in the installed position (Figs. 5 and 6). In dynamic use (Fig. 6), the O-ring should roll in its groove.

IMPORTANT: When installing spool valves, be especially careful of any O-rings. The sharp edges of the spool lands can cut the O-rings unless you are very careful.

O-Ring, Back-Up Ring Application

O-Rings, in combination with Teflon rings, often replace flat and other type of seals in cylinders. An O-ring can be both a static and a dynamic seal when pressure is applied (Fig.4). A back-up ring is sometimes used to keep the O-ring from extruding into the space between the mating parts of the cylinder wall and plunger.

O-ring movement back and forth can create failure if O-rings are installed improperly, if they are the wrong size, or the wrong material for the application. Cylinder wall damage, excessive heat and pressure, and fluid contamination also will damage O-rings.

Checking O-Rings After Installation

Static O-rings which are used as gaskets should be tightened or torqued again after the unit has been warmed up and cycled several times.

Dynamic O-rings should be cycled (moved back and forth through their normal pattern of travel) several times to allow the ring to rotate and assume a neutral position.

All dynamic rings should pass a very small amount of fluid when rotating, which permits a lubricating film to pass between the ring and the shaft. This film prevents scuffing of the ring, which results in short life.

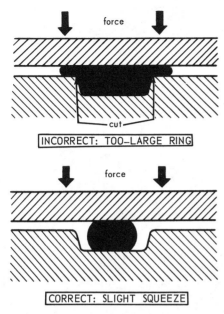

X 1248

Fig. 5 — O-Ring in Static Use

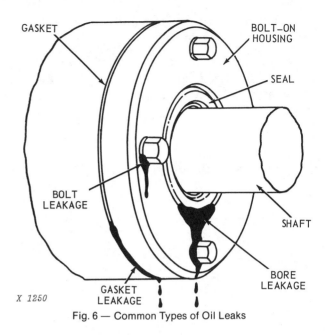

X 1250

Fig. 6 — Common Types of Oil Leaks

MAINTENANCE OF OTHER SEALS

Modern seals use rubber, leather, plastics, and other materials which require special handling. Some of the maintenance rules are given below.

Checking Seals for Leakage

Before disassembling a component, check out the causes of leakage. This may save a return job, caused by problems other than the oil seal.

Before cleaning the area around the seal, find the path of leakage (Fig. 6). Sometimes the leakage may be from sources other than the seal. Leakage could be from worn gaskets, loose bolts, cracked housings, or loose line connections.

Inspect the outside sealing area of the seal to see if it is wet or dry. If wet, see whether the oil is running out or is merely a lubricating film.

Removing Seals

During removal, continue to check for causes of leakage.

Check both the inner and outer parts of the seal for wet oil which means leakage.

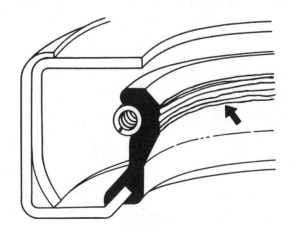

X 1251

Fig. 7 — Seal Worn by Rough Shaft

When removing the seal, inspect the sealing surface or lips (Fig. 7) before washing. Look for unusual wear, warping, cuts and gouges, or particles embedded in the seal.

On spring-loaded lip seals, be sure the spring is seated around the lip, and that the lip was not damaged when first installed.

Do not disassemble the unit any more than necessary to replace the faulty seals.

Checking Shafts and Bores

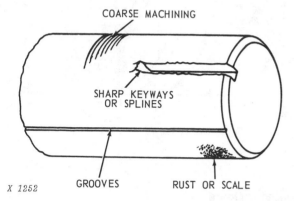

X 1252

Fig. 8 — Shaft Conditions Which Can Damage
Seals and Cause Leakage

Check shafts for roughness at seal contact areas (Fig. 8). Look for deep scratches or nicks that could have damaged the seal.

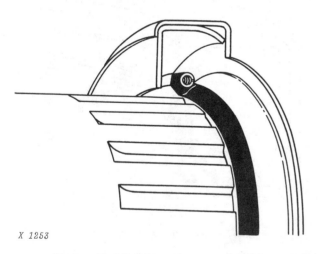

X 1253

Fig. 9 — Shaft Splines or Keyways Can Damage
Seals During Installation

Find out if a shaft spline, keyway, or burred end could have caused a nick or cut in the seal lip during installation (Fig. 9).

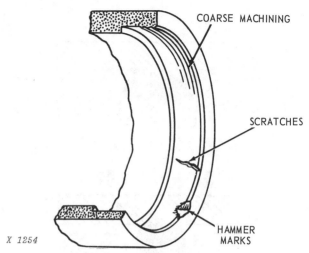

X 1254

Fig. 10 — Bore Conditions Which Can Damage
Seals and Cause Leakage

Inspect the bore into which the seal is pressed (Fig. 10). Look for nicks and gouges that could create a path of oil leakage. A coarsely machined bore can allow oil to seep out by a spiral path. Sharp corners at the bore edges can score the metal case of the seal when it is pressed in. These scores can make a path for oil leakage.

Checking Seals for Compatibility with Fluids or Operating Temperatures

Some hydraulic oils are harmful to certain seals, especially rubber lips. An incorrect oil can either harden or soften the synthetic rubber in seals and so damage them.

If the seal lip is "spongy," this probably means that the seal and the hydraulic fluid are not compatible. If the seal is factory-approved, then an improper fluid has been used in the system (see Chapter 10).

Hardening of the seal lip can be caused by either heat or chemical reaction with an improper fluid.

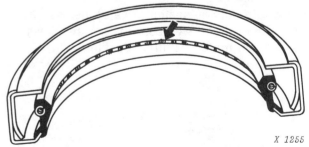

X 1255

Fig. 11 — Seal Lips Damaged by Heat

Hardening of the seal lip on the area of shaft contact (Fig. 11) is generally the result of heat from either the shaft or the fluid.

Installing Seals

1. Install only genuine seals recommended by the manufacturer of the machine.

2. Use only the proper fluids as stated in the machine operator's manual.

3. Keep the seals and fluids clean and free of dirt.

4. Before installing seals, clean the shaft or bore area. Inspect these areas for damage. File or stone away any burrs or bad nicks and polish with a fine emery cloth for a ground finish, then clean the area to remove metal particles. In dynamic applications, the sliding surface for the seal should have a mirror finish for best operation.

5. Lubricate the seal, especially any lips, to ease installation. Use the hydraulic fluid to lubricate the seal. Also soak packings in the hydraulic fluid before installing them.

6. With metal-cased seals, coat the seal's outside diameter with a *thin* film of gasket cement to prevent bore leakage.

NOTE: Pre-coated seals do not require cement on the bore fit.

7. Use a factory-recommended tool to install the seal properly. This is very important with pressed-in seals. If a seal driving tool is not available, 1) use a circular ring such as an old bearing race that contacts the seal case near the outer diameter, or 2) use a square wooden block. *Do not use sharp tools.*

8. Fit packings in snugly without using undue force. Be sure they are not too tight.

9. Use shim stock to protect seals when installing them over sharp edges such as shaft splines. Place rolled plastic shim stock (0.003—0.010 inch) over the sharp edge, then pull it out after the seal is in place.

10. Be sure the seal is driven in evenly to prevent "cocking" of the seal (Fig. 12). A cocked seal allows oil to leak out and dirt to enter as shown. *Be careful not to bend or "dish" the flat metal area*

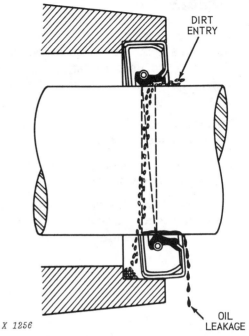

X 1256

Fig. 12 — Cocked Seals Allow Dirt to Enter and Oil to Leak Out

of metal-cased seals. This causes the lips to be distorted.

11. After assembly, always check the unit by hand for free operation if possible before starting up the system.

12. Try to prevent dirt and grit from falling on piston rods, etc. and being carried into the seal. This material can quickly damage the seal or score the metal surfaces.

Run-In Checking of New Lip-Type Seals

When a new lip-type seal is installed on a clean shaft, a break-in period of a few hours is required to seat the seal lip with the shaft surface. During this period, the seal polishes a pattern on the shaft and the shaft in turn seats the lip contact, wearing away the knife-sharp lip contact to a narrow band.

During this period, slight seepage may occur. After seating, the seal should perform without any measurable leakage.

TEST YOURSELF

QUESTIONS

1. (Fill in the blanks with "dynamic" or "static".) "_____ seals are used to seal fixed parts, while _____ seals are used to seal moving parts."

2. What is the most common seal used in farm and industrial hydraulic systems?

3. (True or false?) "Slight leakage is permissible in some dynamic seal applications."

4. (True or false?) "When repairing a component, replace only the seals that are damaged."

HYDRAULIC FLUIDS / CHAPTER 10

INTRODUCTION

The hydraulic fluid is the medium by which power is transmitted from a pump to the mechanisms which produce work such as cylinders and hydraulic motors. The fluid is just as important as any other part of a hydraulic system. In fact, it has been estimated that 70 percent of hydraulic problems stem from the use of improper types of fluids, or fluids containing dirt and other contaminants.

X 1258

Fig. 1—Hydraulic Fluids are Highly Refined Petroleum Oils

When we speak of a hydraulic fluid, in almost all cases we really mean a highly refined petroleum oil (Fig. 1) usually containing additives, some to suppress unwanted properties and others to give the oil desirable properties.

Here it might be well to interject a word of caution. NEVER use hydraulic brake fluid in a hydraulic system designed to use petroleum-base oils. Brake fluid is **not** a petroleum product and is completely incompatible with petroleum-base hydraulic fluids.

During the development of hydraulic equipment, engineers make careful studies of the available fluids to find one best suited to enable their product to give efficient, trouble-free operation. Sometimes it is even necessary to develop a new fluid which has just the right properties. This is why it is always essential to use the fluid recommended in the instructions that accompany a hydraulic machine or mechanism.

WHAT A HYDRAULIC FLUID MUST DO

First of all, of course, a hydraulic fluid must be capable of transmitting the power applied to it. Of equal importance it must do several other things. It must provide lubrication for moving parts, be stable over a long period of time, protect machine parts from rust and corrosion, resist foaming and oxidation, and be capable of separating itself readily from air, water, and other contaminants. The fluid must also maintain proper viscosity through a wide temperature range, and finally, be readily available and reasonably economical to use.

PROPERTIES OF HYDRAULIC FLUIDS
VISCOSITY

For proper power transmission, this is a most important property. Viscosity is a measurement of a fluid's resistance to flow. Said another way, it is a fluid's "thickness" at a given temperature. Viscosity is expressed by SAE (Society of Automotive Engineers) numbers; 5W, 10W, 20W, 30, 40, etc. All petroleum oils tend to become thin as the temperature goes up, and to thicken as the temperature goes down. If viscosity is too low (fluid too thin), the possibility of leakage past seals and from joints is increased. This is particularly true in modern pumps, valves, and motors which depend on close fitting parts for creating and maintaining proper oil pressure. If viscosity is too high (fluid too thick), sluggish operation results and extra horsepower is required to push the fluid through the system. Viscosity also has a definite influence on a fluid's ability to lubricate moving parts.

Viscosity is determined by measuring the time required for 60 cubic centimeters of an oil at a temperature of 210°F. to flow through a small orifice in an instrument known as a Saybolt Viscometer or another instrument called a Kinematic Viscometer (Fig. 2). The actual SAE number is determined by comparing the time required for the oil to pass through the instruments with a chart provided by the Society of Automotive Engineers.

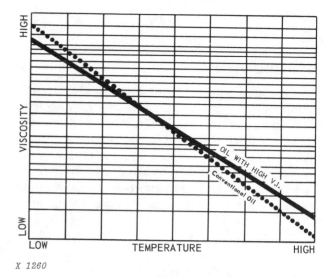

X 1260

Fig. 3—The Viscosity of an Oil with a High VI Contrasted with the Viscosity of a More Conventional Oil

Viscosity Index Improver

Even though carefully refined oils have a good viscosity index, a substance called a viscosity index improver is often added to hydraulic fluids. This substance increases the VI of the fluid so that its viscosity change over a wide range of temperatures is as little as practical.

WEAR PREVENTION

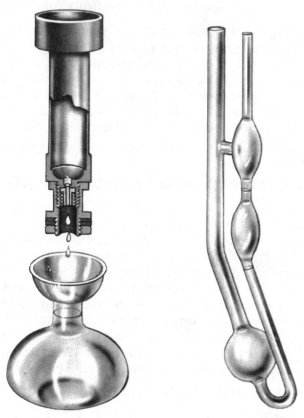

Fig. 2 — Saybolt Viscometer (Left) and Kinematic Viscometer (Right)

Viscosity Index (VI)

This is simply a measure of a fluid's change in thickness with respect to changes in temperature. If a fluid becomes thick at low temperatures and very thin at high temperatures, it has a **low** VI. On the other hand if viscosity remains relatively the same at varying temperatures, the fluid has a **high** VI (Fig. 3). As pointed out earlier, in a fluid with good viscosity characteristics, there is a balance between a fluid thick enough to prevent leakage and provide good lubrication while, at the same time, being thin enough to flow readily through the system. Therefore, a fluid with a high VI is almost always desirable and must be an important consideration in hydraulic fluid recommendations.

X 1196

Fig. 4—Vane Pump Ring Worn Due to Lack of Lubrication

Most hydraulic components are fitted with great precision, yet they must be lubricated to prevent wear (Fig. 4). To provide this lubrication, a satisfactory fluid must have good "oiliness" to get in the tiny spaces available between moving parts and hold friction between the parts to a minimum.

A good fluid must also have the ability to "stick" to these closely fitted parts, even when they are quite warm.

Good lubricating qualities become even more important in many modern machines where the hydraulic fluid has a double purpose—to operate the hydraulic mechanisms, and also to lubricate the transmission, differential, and other parts of the machine.

The best hydraulic fluids contain an "extreme pressure" additive which assures good lubrication of very close fitting metal-to-metal parts operating at high pressures and temperatures. This additive reduces friction and helps to prevent galling, scoring, seizure, and wear.

RESISTANCE TO OXIDATION

X 1191

Fig. 5—Radial Pump Pistons Scored by Contaminated Fluid

X 1193

Fig. 6—Vane Pump Rotor Ring Worn and Pitted by Contaminated Fluid

Everyone is familiar with the effects of air on a piece of shiny iron, especially in the presence of water; it combines with oxygen in the air to form rust and other foreign materials. Its chemical properties are also changed. Just like the iron, all oils combine to some extent with the oxygen in the air. This changes the oil's chemical composition. Organic acids are formed which may be harmful to metal parts (Figs. 5 and 6) and many types of seals and packings in the system. In addition to the acids, sludges are often formed through the reactions between fluids and air. Both reactions are speeded up by the presence of water and other contaminants such as dust, dirt and metallic particles in the fluid. This is one reason why an efficient filtering system is so important in any hydraulic system. Heat is also a very important factor in oxidation. It has been determined, for instance, that for every 18°F. rise in temperature, the rate of oxidation doubles. Because of this, some hydraulic systems contain a cooler to hold temperatures to reasonable limits.

Fortunately, carefully refined fluids, plus the addition of a special chemical, successfully resist oxidation. With careful attention to prevent the entrance of dirt and other contaminants, most modern hydraulic fluids will operate for many hours without ill effects due to oxidation. However, oxidation can be a real problem unless high-quality hydraulic fluids, specifically recommended by the equipment manufacturer are used.

RUST AND CORROSION PREVENTION

Rust and corrosion are both related to oxidation, and a hydraulic fluid (providing it is kept clean) with good anti-oxidation qualities is likely to resist rust and corrosion. However, the possibilities of rust or corrosion developing are always present and they cannot be ignored. Rust and corrosion differ in that rusting adds to metal making the part larger while corrosion, caused by acids or local electrochemical cells, is an eating away of the metal. Either condition, of course, is highly detrimental to a hydraulic mechanism. Rust causes rough spots (Fig. 7) which damage seals and close fitting parts. It is most likely to form during storage periods, down-time, or even overnight. Corrosion affects the fit of closely machined parts and permits undesirable leakage. Both rust and corrosion cause erratic operation and untimely wear.

X 1194

Fig. 7—Pump Drive Cam Pitted by Rust

Good hydraulic fluids contain both rust and corrosion additives which neutralize corrosion-forming acids and cling to metal parts to protect them from rusting and corroding.

RESISTANCE TO FOAMING

The proper operation of any hydraulic system is based on the fact that fluids cannot be compressed by pressures normally encountered in the system. In effect the fluid acts like a "liquid" steel rod. Any force applied to it at one end is transmitted to the other end without any "slack" due to compression. However, air which **is** compressible can be absorbed by the fluid. In many systems the fluid reservoir is directly exposed to the atmosphere which promotes entrance of air into the fluid. In addition, air can enter the system through defective packings, leaky lines, or if the fluid level in the reservoir is allowed to get too low. In many systems turbulence promotes the mixing of fluid and air.

Good fluids have the capacity to "dissolve" a small amount of air. The amount of air that can be dissolved increases as pressure and temperature increase. This dissolved air has no harmful effects on operation. But if the amount of air which enters the fluid is greater than the fluid's capacity to dissolve it, bubbles form which, since air is compressible, result in mushy, unsatisfactory operation. Furthermore, some air in solution under pressure comes out of solution when pressure is released. This air creates foam which seriously affects proper action, and especially, lubrication.

While most well refined oils are not subject to excessive foaming, most good hydraulic fluids contain a foam inhibitor additive which speeds up the rate at which bubbles break up. This improves the fluid's ability to do its work properly, and to

increase its capacity for adequately lubricating moving parts.

ABILITY TO SEPARATE FROM WATER

Contrary to popular opinion, oil and water **will** mix. This mixture is called an "emulsification." It is almost impossible to keep all water out of a hydraulic system. Water vapor enters the reservoir where it condenses into droplets. It also may enter through tiny leaks in the system. Because of the violent agitation, churning, and continual recirculation in a typical hydraulic system, the water and fluid quickly mix to form an emulsion. Any appreciable amount of water in the fluid is highly detrimental. The emulsion promotes rust (Fig. 8), increases oxidation which forms acids and sludges, and reduces the fluid's ability to lubricate moving parts properly. Also, emulsions often have a slimy, sticky, or pasty consistency which interferes with normal operation of valves and other parts.

X 1195

Fig. 8—Balanced Vane Pump Rusted by Water in Fluid

LACK OF OTHER CONTAMINANTS

It should go without saying that a good hydraulic fluid should be as free as possible of contaminants such as metallic particles, dust, dirt and the like. Such materials are not only likely to damage closely fitted parts, but they also seem to help the undesirable oxidation process to take place.

Use of reliable fluids, careful storage, good filters, proper handling of fluids, and periodic cleaning of hydraulic systems all materially reduce the danger of contamination.

MAINTAINING GOOD FLUID

As has been said many times, dirt and contamination are the worst enemies of any hydraulic system. Continued, long operation at high efficiency is very dependent upon proper fluid maintenance. First of all, **only** a fluid recommended by the

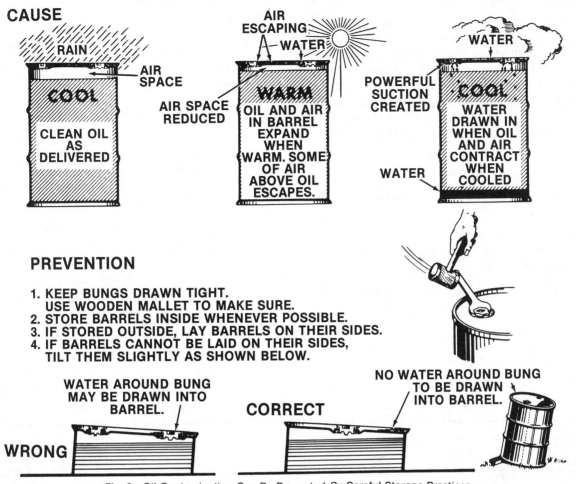

CAUSE

RAIN

AIR SPACE

COOL CLEAN OIL AS DELIVERED

AIR ESCAPING — WATER

AIR SPACE REDUCED

WARM OIL AND AIR IN BARREL EXPAND WHEN WARM. SOME OF AIR ABOVE OIL ESCAPES.

WATER

POWERFUL SUCTION CREATED

WATER

COOL WATER DRAWN IN WHEN OIL AND AIR CONTRACT WHEN COOLED

PREVENTION

1. KEEP BUNGS DRAWN TIGHT. USE WOODEN MALLET TO MAKE SURE.
2. STORE BARRELS INSIDE WHENEVER POSSIBLE.
3. IF STORED OUTSIDE, LAY BARRELS ON THEIR SIDES.
4. IF BARRELS CANNOT BE LAID ON THEIR SIDES, TILT THEM SLIGHTLY AS SHOWN BELOW.

WATER AROUND BUNG MAY BE DRAWN INTO BARREL.

WRONG

CORRECT

NO WATER AROUND BUNG TO BE DRAWN INTO BARREL.

Fig. 9—Oil Contamination Can Be Prevented By Careful Storage Practices

manufacturer of the system should be used; it should be checked at the suggested intervals maintained at the correct level, properly filtered, and changed at the recommended intervals.

DRAIN SCHEDULE

Periodic drainage of the entire system is very important. This is the only way to remove contaminants, products of oxidation such as sludge and acids, and other particles that may be injurious to the system. Actually, in modern hydraulic systems using approved fluids, the drain period is not frequent and it should work no hardship on the part of the owner to follow the manufacturer's directions. Many modern fluids are so highly refined, filtered, and fortified by additives that system flushing is not necessary. However if flushing is recommended by the manufacturer, it is always advisable to follow directions so as not to contaminate the new oil with flushing oil that cannot be drained from the system.

See Chapter 11 for details on draining, cleaning, and flushing systems.

KEEPING HYDRAULIC FLUIDS CLEAN

All good hydraulic fluids, which come in cans or barrels, are delivered perfectly clean and free from contaminants. It is when the containers are opened or stored that troubles develop.

When opening a can or barrel, be absolutely certain that the area around the opening is completely free of dust, dirt, lint, or water. If a container, funnel, or hose is required to fill the system, be sure that it is spotless.

When possible, always store barrels of hydraulic fluid indoors, or at least under cover, and be sure the bung is tight. If barrels are stored in the sun without a tight bung, the fluid will expand, forcing some air from the bung. Then as the fluid cools off, the fluid contracts, drawing any rain, dew, or other moisture into the barrel and fluid. The detrimental effects of water in hydraulic fluids have already been discussed. For this reason, keep the bungs in barrels as tight as possible, and tip the barrels in such a manner that water cannot collect around the bung (Fig. 9).

CHOOSE OIL WITH THE RIGHT VISCOSITY

Choosing a quality oil with the right viscosity will extend component life and help to ensure troublefree operation. Manufacturers conduct extensive performance tests on how specific oils function under different operating and climatic conditions in their equipment. So, always follow the manufacturer's recommendations. Be particularly attentive to the oil recommendations for different temperature ranges. See Fig. 10.

NOTE: The hydraulic system reservoir also supplies the oil for the steering system.

Engine oil may be used provided it meets one or more of the following:

API Service CD/SF, CD/SE, CD/SD, CD/SC, CC/SF, CC/SE, CC/SD, CC/SC, (MIL-L-2104C, MIL-L-2104D, MIL-L-46152B).

Oil meeting MIL-L-46167A may be used as an arctic oil.

TEST YOURSELF
QUESTIONS

1. What are good hydraulic fluids composed of?

2. At higher temperature do oils become thicker or thinner?

3. Is a high-viscosity oil thicker or thinner than a low-viscosity oil?

4. What happens when a hydraulic fluid oxidizes?

5. Why is it important to follow manufacturer's recommendations?

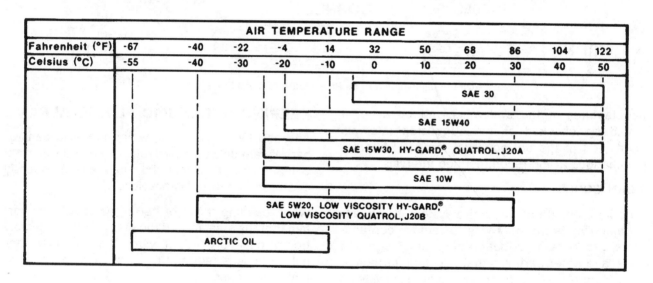

Fig. 10 — Typical Oil Recommendation

INTRODUCTION

A hydraulic system is fairly easy to maintain: the fluid provides a lubricant and protects against overload. But like any other mechanism, it must be operated properly. You can damage a hydraulic system by too much speed, too much heat, too much pressure, or too much contamination.

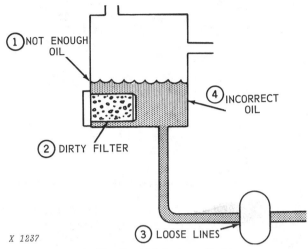

Fig. 2—The Key Maintenance Problems

X7626

Fig. 1—The Machine Will Respond to Good Maintenance

Proper maintenance will reduce your hydraulic troubles. By caring for the system using a regular maintenance program, you can eliminate common problems and anticipate special ones. These problems can then be corrected before a breakdown occurs.

Other chapters in this manual tell you how to diagnose failures and remedy them. This chapter will explain how to keep the system going while it is in operation.

Here are the key maintenance problems:

1. *Not enough oil in the reservoir.*

2. *Clogged or dirty oil filters.*

3. *Loose intake lines.*

4. *Incorrect oil in the system.*

All of these problems can be solved or prevented by knowing the system and maintaining it properly.

Let's discuss some of the practices which will keep the hydraulic system in top-notch condition.

 CAUTION: Before any maintenance or service is performed, make sure that:

- **Machine is shut off**

- **Hydraulic pressure is relieved**

- **All equipment is lowered to the ground**

MAINTENANCE OF THE WHOLE SYSTEM

This section covers the general maintenance which keeps the whole system at peak performance.

THE IMPORTANCE OF CLEANLINESS

Cleanliness is No. 1 when it comes to servicing hydraulic systems. KEEP DIRT AND OTHER CONTAMINANTS OUT OF THE SYSTEM! Small particles can score valves, seize pumps, clog orifices and so cause expensive repair jobs.

How do you keep the hydraulic system clean? Let's put it this way:

- **Keep the oil clean**

- **Keep the system clean**

- **Keep your work area clean**

- **Be careful when you change or add oil**

In detail, here's how these steps work.

KEEP THE OIL CLEAN

X 1238

Fig. 3—Keep the Oil Clean

Strive to keep oil clean from the minute it is delivered to you. Choose a clean location for storing the oil. When the oil is taken out of storage, use only clean containers with lids for carrying oil from storage to point of use. Use a clean funnel fitted with a fine-mesh screen when pouring oil from the container into the reservoir.

Keep an adequate supply of clean strainers, funnels, and oil containers. Store them in a clean, dust-free cabinet. Use a clean, lint-free cloth to wipe the dipstick when checking the oil level.

Caution machine operators to do everything possible to keep dirt from getting into the oil during operation of the machine.

KEEP THE SYSTEM CLEAN

Change the hydraulic oil and oil filters regularly (Fig. 4). Wipe away dirt and grease before removing filler caps or dipsticks. Steam clean or use solvents to clean the machine areas before removing hydraulic components.

NOTE: When steam cleaning or using water to clean a machine, be sure that filler openings, breather caps, etc. are protected from possible entry of water into the system.

Use clean plastic plugs or caps to cover ends of disconnected lines, or to plug openings when working on a hydraulic system.

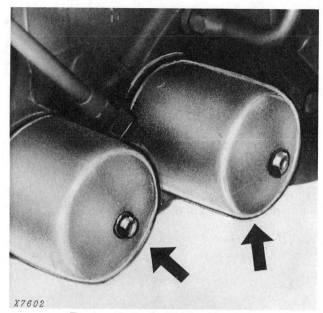

X7602

Fig. 4 — Hydraulic Oil Filter Housings

When removing parts for service, clean them with a suitable solvent, and store them in plastic bags or other clean containers until they are installed again.

When cleaning hydraulic parts, use extreme care to prevent harm to closely fitted, finely finished parts. Use gum solvents or chemical cleaners to clean **metal** parts only. Do not allow these cleaners and solvents to come in contact with seals or gaskets.

Thoroughly rinse the cleaned parts and dry them using compressed air. Protect the parts immediately with a coating of rust preventive oil.

KEEP YOUR WORK AREA CLEAN

A clean work bench is an absolute must when servicing hydraulic components (Fig. 5).

An industrial-type vacuum cleaner is a valuable aid in removing dust, dirt, and tiny metal particles from the work area.

Check the condition of the tools you use—they should be clean. Always use hammers made of plastic, leather, or brass so there is no danger of metal chips getting into components.

THIS...   ...OR THIS?

X 1239

Fig. 5—Keep Your Work Area Clean

BE CAREFUL WHEN YOU CHANGE OR ADD OIL

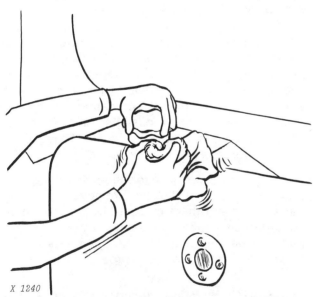

X 1240

Fig. 6—Don't Add Dirt to the System—Just Oil

When checking oil level or adding oil to the system, be sure to clean the areas around the dipstick and filler cap before removing them (Fig. 6).

Before adding oil be sure the oil still in the system is clean. If not, drain the oil and refill the system completely with new oil that will give good performance under existing conditions.

However, if the drained oil has sludge, sediment, or gum and lacquer formations, the system should be cleaned and flushed before refilling.

(Details on cleaning and flushing are given later in this chapter.)

IMPORTANCE OF OIL AND FILTER CHANGES

You can't get peak performance out of a hydraulic system that isn't clean.

Despite all the precautions you take when working with the hydraulic system, some contaminants will get into the system anyway. Good hydraulic oils will hold these contaminants in suspension and the filters will collect them as the oil passes through. A good hydraulic oil contains many additives (see Chapter 10) which work to keep contaminants from damaging or plugging the system. However, these additives lose their effectiveness after a period of time.

Therefore, change the oil at the recommended intervals to make sure the additives will do their job.

The system filters can absorb only a limited amount of dirt particles and other contaminants from the oil. After that the filters stop working.

At this point, clean the filters or replace them with new ones so the cleaning process can be maintained.

DRAINING THE SYSTEM

Periodic draining of the entire hydraulic system is very important. This is the only positive way to completely remove contaminants, oxidized fluid, and other injurious substances from the system.

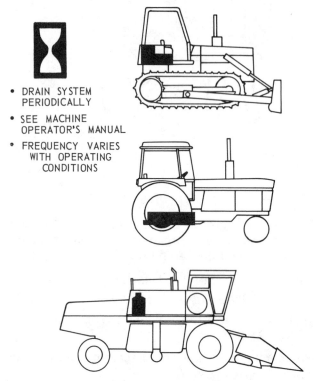

- DRAIN SYSTEM PERIODICALLY
- SEE MACHINE OPERATOR'S MANUAL
- FREQUENCY VARIES WITH OPERATING CONDITIONS

Fig. 7—Drain the System Regularly

The frequency of draining depends on such things as the temperature of operation and the severity of working conditions. The drain schedule recommended by the manufacturer of the equipment should be maintained. The machine operator's manual will tell the method to be used, and the frequency, depending on conditions.

CLEANING AND FLUSHING THE SYSTEM

The nature and amount of deposits in a particular system may vary widely. Inspection may show any condition between a sticky, oily film and a hard, solid deposit (gum or lacquer formation) which completely chokes small oil passages.

If the system is drained frequently enough, the formation of gum and lacquer will be reduced.

When no gum or lacquer formation is suspected, clean the system as follows: After draining the system, clean any sediment from the reservoir, and clean or replace the filter elements.

It may be advisable to flush out the old oil remaining in the system after draining, particularly if the oil is badly contaminated. For this flushing, use the hydraulic fluid recommended for the system involved.

Operate the equipment to cycle the flushing oil through the system. It is important that the valves be manipulated so that the new oil goes through all lines. The time necessary to clean the system will vary, depending on the condition of the system. Run the oil through the system until inspection shows the equipment to be in satisfactory condition, or until it is obvious that the system will have to be disassembled and cleaned manually. Usually from 4 to 48 hours is sufficient for most systems.

Drain out the flushing oil and refill the system with clean hydraulic oil of the recommended type. Be sure to clean or replace the system filters before refilling the system.

NOTE: Most solvents and chemical cleaners on the market today are NOT recommended for use in flushing hydraulic systems: 1) They are poor lubricants, resulting in damage to moving parts, especially the pump. 2) They are difficult to remove completely from the system. Just a trace of some of the commercial chlorinated solvents may be enough to break down the oxidation resistance of even the best hydraulic oils. Also, in the presence of a small amount of water, some of these solvents will corrode steel and copper.

If gums or lacquers have formed on working parts, and the parts are "sticking," remove the affected parts and clean them thoroughly. Do the same if solid particles such as packing material or metal particles have gotten into the system as a result of wear or damage to the components.

 CAUTION: **Before disconnecting parts of the system, relieve all hydraulic pressure by cycling the control levers. Also discharge the accumulator, if used (see Chapter 6).**

When cleaning the disassembled parts, be extremely careful to prevent harm to the closely fitted, finely finished parts. Use of gum solvent or other non-corrosive chemical cleaner is permissible on metal parts ONLY. **Do not allow these materials to come in contact with seals and packings.**

Rinse each cleaned part thoroughly, blow dry with compressed air, and then coat it immediately with a hydraulic oil containing a rust inhibitor. Normally, use the same oil as in the system for this purpose. Freshly cleaned metal surfaces may rust quickly, so be sure they are protected from rust.

After all parts have been cleaned, reassemble the system. Use care to see that no dirt, lint, pipe-thread compound, etc., gets into the system.

As a final step, flush out the system as described above.

FILLING THE SYSTEM

Before filling the system, be sure the area around the filler cap is clean. Fill the reservoir to the specified level with the recommended hydraulic oil. Use only clean oil and funnels or containers. Then be sure to replace the filler cap before operating the equipment.

X7626

Fig. 8—The Machine Will Respond to the Proper Lubrication

Start the engine and warm up the hydraulic system. Then run the equipment through its working cycle at least four times to bleed air from the system.

Add oil if necessary, and operate the machine until the equipment will function smoothly at its full, rated capacity.

With the equipment at rest and the engine shut off, recheck the oil level. If necessary, add oil to bring oil to the proper level.

IMPORTANT: Always check the oil level after any repairs have been made on the system.

PREVENTING LEAKS

What causes leaks? There are hundreds of causes, but they fall into two basic types:

• **Internal Leakage**

• **External Leakage**

Internal leakage does not result in actual loss of oil, but it does reduce the efficiency of the system. External leakage does result in direct loss of oil and can have other undesirable effects as well.

INTERNAL LEAKAGE

Internal leakage as a thin oil film is built into the working parts of a hydraulic system. This lubricates the mating surfaces of valve spools, cylinder pistons, and other moving parts. Oil is not lost through this normal internal leakage since it eventually returns to the system reservoir.

However, too much internal leakage will slow the operation of the system and waste power through the generation of heat. In some cases, it may cause cylinders to creep or drift. Or it may cause loss of oil control in the valves.

Internal leakage increases with the normal wear of parts. Leakage is accelerated by using oil which has too low a viscosity because this oil thins faster at higher temperatures. High pressures also force more oil out of leaking points in the system. This is one reason why excessive pressures can actually reduce the efficiency of the hydraulic system.

Internal leaks are hard to detect. All you can do is watch the operation of the system for sluggish action or creeping and drifting. When these signs appear, it's time to test the system and pinpoint the trouble (see Chapter 12).

EXTERNAL LEAKAGE

External oil leaks not only look bad—they can be expensive and hazardous. A drop of oil every second from a leaking connector can cost the operator quite a lot of money.

A small leak can also be the signal for a hydraulic rupture that may injure a person while putting the machine out of operation.

Every joint in a hydraulic circuit is a potential point of leakage. This is why the number of connections in a system is kept at a minimum.

Components can leak, but care in assembly and use of new seals and gaskets during overhaul will help to reduce this problem.

Fig. 9—Checking Oil Lines for Leaks

⚠ **CAUTION: Escaping fluid under pressure can penetrate the skin causing serious injury. Relieve pressure before disconnecting hydraulic or other lines. Tighten all connections before applying pressure. Keep hands and body away from pinholes and nozzles which eject fluids under high pressure. Use a piece of cardboard or paper to search for leaks. Do not use your hand.**

If ANY fluid is injected into the skin, it must be surgically removed within a few hours by a doctor familiar with this type injury or gangrene may result.

The lines that connect the different parts of the system are the No. 1 source of external leaks. Proper use and care of hoses, tubes, and pipes is covered in detail in Chapter 8. Here are a few key points:

1) If the reservoir oil level is lower than normal, check all external oil lines for leaks.

2) Pin-hole leaks are hard to detect, yet they can be dangerous. A "vapor" of oil from a small leak can create a fire hazard, or a fine spray of oil against a hot engine can ignite.

3) The rubber cover on flexible hoses may crack or split without actually leaking. But check very closely for internal damage. The depth of the crack is the deciding factor. Any oil dampness is a sign the hose is leaking.

4) Air leaks in suction lines are hard to locate. One way is to pour oil over the points where you suspect leaks. If the noise or bubbling in the system stops, you've located the leak.

5) Leaks in concealed lines are also hard to find. One routine test for open-center systems is to install a pressure gauge in the discharge line near the pump and then to block off the circuit in progressive stages. When the gauge shows a pressure

drop at a given blocking point, the leak is between this point and the one blocked just before it.

NOTE: When making Step 5 test on open-center systems, be careful not to deadend the pump as serious damage to the pump and lines could result.

6. If line connections are leaking, *tighten only until the leak stops.* If the connection will not stay tightened, the threads are probably stripped and the connector must be replaced. If the connector will tighten but still leaks, check for a cracked line flare or a damaged seal. But remember: *More damage has been done to line connectors by overtightening than from any other cause.*

Other chapters in this manual give details on checking and preventing leaks in components such as pumps (Chapter 2), valves (Chapter 3), and cylinders (Chapter 4).

After stopping leaks in a system, be sure to warm up the system and cycle the equipment, then recheck the trouble spots to be sure the leaks are stopped.

Recheck the system oil level and replace any oil lost through leaks or broken connections.

PREVENTING OVERHEATING

Heat causes hydraulic oil to break down faster and lose its effectiveness. This is why cooling of the oil is needed.

In many systems, enough heat is dissipated through the lines, the components, and the reservoir to keep the oil fairly cool. But on high-pressure, high-speed circuits, oil coolers are needed to dissipate the extra heat.

Overheating the system can:

• **Break down the oil**

• **Damage the seals**

• **Coat parts with varnish deposits**

• **Cause extra leakage past working parts**

• **Reduce the output of the system**

To help prevent overheating, keep the oil at the proper level, clean dirt and mud from lines, reservoirs, and coolers, check for dented and kinked lines, and keep relief valves adjusted properly. Also be careful not to overspeed or overload the system and never hold control valves in power position longer than necessary.

If the system still overheats, refer to the charts at the end of Chapter 12 which list the causes and remedies for overheating.

THERMAL (HEAT) EXPANSION

 CAUTION: Thermal expansion can raise system pressure dramatically. Be careful around stored machines that are exposed to heat.

Thermal expansion is simply the expanding of oil in a system at rest due to heat. This expansion raises the pressure in the system. One degree rise in temperature may cause a pressure rise of 50 to 60 psi in a tightly blocked system.

Normally a hydraulic circuit has built-in escape hatches for this heat expansion. (During operation, the oil circulates and the system adjusts itself to heat changes.) But if there is no outlet, as during storage, a part of the system can "rupture."

In some hydraulic cylinders, for example, there is limited room for expansion, and the heat from the sun could cause breakage.

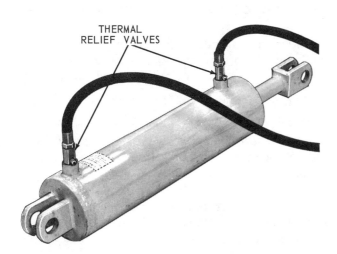

Fig. 10—Thermal Relief Valves on a Hydraulic Cylinder

Thermal relief valves (Fig. 10) are used to combat this problem. Another solution is for the operator to make sure that the system is not blocked off at any time while it is at rest. Cylinders which have no thermal relief valves should be partly drained before being put into storage.

PREVENTING AIR-IN-OIL PROBLEMS

If air gets into the system, it can result in 1) "spongy" action of equipment, 2) chattering in the system, 3) a noisy pump, and 4) a pump not operating.

If the oil level in the reservoir is too low, air bubbles will form in the reservoir. Air can also get into the system through leaks in the suction line when the oil lines are opened to make repairs, or when the system is drained and refilled.

To keep air out of the system:

1. Be sure the oil in the reservoir is kept at the correct level.

2. Replace any leaking sections of the suction line.

3. Tighten any connections that are leaking. Tighten only until the leak (or noise) stops.

4. After making any repairs and refilling system, cycle the equipment at least four times to bleed all air from the system. (Be sure to recheck reservoir oil level after cycling.) Bleeding may also improve the operation of new machines after a few hours of use.

5. When attaching remote cylinders to the system, bleed air from them as outlined at the end of Chapter 4.

WHAT IS GOOD MAINTENANCE?

- **Use Common Sense**
- **Stop, Look, Touch, Listen Before Any Disassembly Is Started**
- **Keep Parts Clean**
- **Change Oil and Filters Regularly**
- **Maintain Good Records**

GOOD GUYS	BAD GUYS
1. Cleanliness	1. Dirt
2. High Quality Oil	2. Water
3. Proper Filters	3. Air
4. Tight Seals	4. Heat
5. Normal Operation	5. Abuse

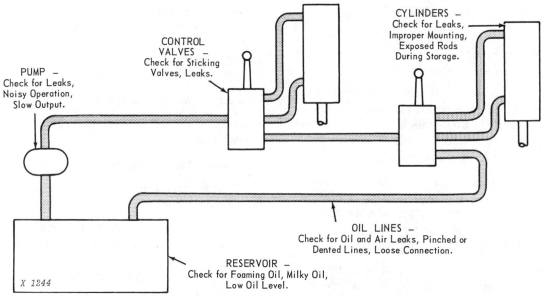

PUMP —
Check for Leaks,
Noisy Operation,
Slow Output.

CONTROL
VALVES —
Check for Sticking
Valves, Leaks.

CYLINDERS —
Check for Leaks,
Improper Mounting,
Exposed Rods
During Storage.

OIL LINES —
Check for Oil and Air Leaks, Pinched or
Dented Lines, Loose Connection.

RESERVOIR —
Check for Foaming Oil, Milky Oil,
Low Oil Level.

X 1244

Fig. 11—Check the Whole System Before Operation

CHECKING THE SYSTEM BEFORE OPERATION

After repairs on the system, check the whole circuit for leaks, proper oil level, overheating, etc. *Do this before operating the system on the job.*

For accurate checks, warm up the system and cycle the hydraulic equipment as described under "Filling the System."

Also check the system periodically during operation. Follow the machine operator's manual for the proper intervals.

Following are some areas which should be checked regularly.

CHECKING RESERVOIR AND OIL

1. Check the reservoir oil level and the condition of the oil frequently.

2. At this time, also check for other malfunctions in the system. Look for:

   a) *Oil Bubbles or Foaming Oil.* This may mean an air leak somewhere in the system.

   b) *Change in Oil Level.* If there is a noticeable change in the oil level from day to day, look for leaks or cracks in the external parts of the system.

   c) *Milky Oil.* This indicates water in the system or in the oil used. Be sure that oil is stored properly (see Chapter 10).

3. Before removing the filler cap, wipe away all dirt and grime. If a dipstick is used for checking oil, wipe it with a clean, lint-free cloth.

CHECKING COOLER, LINES AND CONNECTORS

1. Clean the oil cooler periodically and check it for leaks. Keep the fins clean on air-to-oil coolers. Check for corrosion in water-to-oil types.

2. Check oil lines and connectors at regular intervals. Look for:

   a) *Pressure Oil Leaks.* Oil leaks in the pressure side of the system can be located by examining the outside of lines and fittings.

 CAUTION: Never try to locate a leak by running your hand over the suspected area. Always use a piece of cardboard.

   b) *Air Leaks.* If the suction side of the system is drawing in air, the oil in the reservoir will bubble and foam.

   c) *Pinched or Dented Lines.* This can cause oil foaming, overheating, and loss of hydraulic power. Replace damaged hoses or tubes at once. Wash lines inside and out with clean solvent before installing them.

3. Tighten any loose lines or connections. Use two wrenches to avoid twisting hose or tubes. Replace any connectors that continue to leak.

IMPORTANT: Tighten loose connectors ONLY until the leak stops.

CHECKING VALVES

Dirt will cause valves to stick or work erratically. Also, after long use, valve spools may become worn, allowing oil to leak past them. Check all valves for leaks periodically. Service the valves as explained in Chapter 3.

CHECKING CYLINDERS

Check cylinders periodically for both external and internal leakage. Make sure cylinders are mounted properly. Be sure the rods of exposed cylinders are not left extended when the machine is stored. Otherwise the rods will collect dirt and moisture which may cause rusting or pitting of the rods. If the rods must be exposed, make sure they are coated with a heavy grease. See Chapter 4 for more details on cylinder maintenance.

CHECKING PUMPS

Check external pumps for possible leaks at mating surfaces of the housing and at cap screws. Check the pump with the machine running to see if it is moving the hydraulic equipment at a satisfactory speed. If not, make the pump and system tests given in Chapter 12 to locate the trouble.

CHECKING MOTORS

Never permit a hydraulic motor to overheat. If it is running hot, make sure the oil supply is adequate and also check the system oil cooler to make sure it is functioning properly. Also inspect for leaks at the motor hose connections, around the shaft, at the seals, and at mating surfaces.

WHEN MAINTENANCE FAILS . . .

In this chapter, we have covered the maintenance which helps to *prevent* failures of the system.

However, with the best of care, hydraulic components will fail at times. When this happens, you must first locate the problem. This is the job of another chapter which follows, Chapter 12 on "Diagnosis and Testing of Hydraulic Systems."

TEST YOURSELF
QUESTIONS

1. What four maintenance problems do servicemen discover most often when answering a customer's complaints?

2. True or false? "When flushing the hydraulic system, use a nonpetroleum solvent or a chemical cleaner."

3. (Fill in the blank.) Thermal expansion is the expanding of oil due to _____.

4. Match the items in the left-hand column with the likely cause in the right-hand column:

a. Foaming Oil   1. Water

b. Milky Oil   2. Heat

c. Scorched Oil   3. Air

DIAGNOSIS AND TESTING
OF HYDRAULIC SYSTEMS / CHAPTER 12

INTRODUCTION

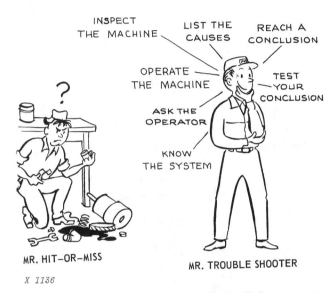

X 1136

Fig. 1—Which Would You Rather Be?

Mr. Hit-or-Miss or Mr. Trouble Shooter—which would you rather be? Both have the title of serviceman but don't be fooled by that.

Mr. Hit-or-Miss is a parts exchanger who dives into a machine and starts replacing parts helter-skelter until he finds the trouble—maybe—after wasting a lot of the customer's time and money.

Mr. Trouble Shooter starts out by using his brain. He gets all the facts and examines them until he has pin-pointed the trouble. Then he checks out his diagnosis by testing it and *only then* does he start replacing parts.

Mr. Hit-or-Miss is fast becoming a man of the past. What dealer can afford to keep him around at today's prices?

With the complex systems of today, diagnosis and testing by Mr. Trouble Shooter is the only way.

SEVEN BASIC STEPS

A good program of diagnosis and testing has seven basic steps:

1. **Know the System**
2. **Ask the Operator**
3. **Operate the Machine**
4. **Inspect the Machine**
5. **List the Possible Causes**
6. **Reach a Conclusion**
7. **Test Your Conclusion**

Let's see what these steps mean.

1. KNOW THE SYSTEM

X 1138

Fig. 2—Know the System

In other words, "Do your homework." Study the machine technical manuals. Know how the system works, whether it's open- or closed-center, what the valve settings and pump output should be.

Keep up with the latest service bulletins. Read them and then file in a handy place. The problem on your latest machine may be in this month's bulletin, giving the cause and remedy.

You can be prepared for any problem by knowing the system.

2. ASK THE OPERATOR

X7598

Fig. 3—Ask the Operator

A good reporter gets the full story from a witness—the operator.

He can tell you how the machine acted when it started to fail, what was unusual about it.

Try to find out too if any do-it-yourself service was performed. (You may find out later that the trouble is somewhere else. But you should know if any valves were tampered with, etc.)

Ask about how the machine is used, when it is serviced. Many problems can be traced to poor maintenance or abuse of the machine.

3. OPERATE THE MACHINE

X7599

Fig. 4—Operate the Machine

Get on the machine and operate it. Warm it up and put it through its paces. Always verify the operator's story — check it out yourself.

Are the gauges reading normal? (If not, it may be hydraulic trouble or it may mean the gauges are faulty.)

How's the performance? Is the action slow, erratic, or nil?

Do the controls feel solid or "spongy"? Do they seem to be "sticking"?

Smell anything? Any signs of smoke?

Hear any funny sounds? Where? At what speeds or during what cycles?

4. INSPECT THE MACHINE

X7628

Fig. 5—Inspect the Machine

Now get off the machine and make a visual check. Use your eyes, ears, and nose to spot any signs of trouble.

First inspect the oil in the reservoir. How is the oil level? Is the oil foamy? Milky? Does it smell scorched? Does it appear to be too thin or too thick? How dirty is it?

If the oil is very dirty, also check the filters for clogging.

Feel the reservoir. Is it hotter than normal? Is it caked with dirt and mud? Is the paint peeled off from heat? Check the pump inlet line for restrictions. Check for collapsed hoses.

⚠ CAUTION: Do not feel for pinhole leaks. Escaping fluid under pressure can penetrate the skin causing injury. Relieve all hydraulic pressure before working on a pressurized hydraulic line or component.

Follow the circuit and keep on checking. Look for oil leaks at line connectors. Watch for air leaks at loose clamps, etc.

Check the oil cooler. Is it free of trash and mud?

Look closely at the components. Inspect for cracked welds, hairline cracks, loose tie bolts, or damaged linkages.

While you inspect, make a note of all the trouble signs.

5. LIST THE POSSIBLE CAUSES

X7629

Fig. 6—List the Possible Causes

Now you are ready to make a list of the possible causes.

What were the signs you found while inspecting the machine? And what is the most likely cause?

Are there other possibilities? Remember that one failure often leads to another.

6. REACH A CONCLUSION

Look over your list of possible causes and decide which are most likely and which are easiest to verify.

Use the Trouble Shooting Charts at the end of this chapter as a guide.

Reach your decision on the leading causes and plan to check them first.

X7600

Fig. 7—Reach a Conclusion

7. TEST YOUR CONCLUSION

X7601

Fig. 8—Test Your Conclusion

Now for the final step: Before you start repairing the system, test your conclusions to see if they are correct.

Some of the items on your list can be verified without further testing. Analyze the information you already have:

Were all the hydraulic functions bad? Then probably the failure is in a component that is common to all parts of the system. Examples: pump, filters, system relief valves.

Was only one circuit bad? Then you can eliminate the system components and concentrate on the parts of that one circuit.

Now your list is beginning to narrow so that you can point your tests at one or two components.

The next part of this chapter will tell you how to test the system and pinpoint these final troubles.

But first let's repeat the seven rules for good trouble shooting:

1. **Know the System**
2. **Ask the Operator**
3. **Operate the Machine**
4. **Inspect the Machine**
5. **List the Possible Causes**
6. **Reach a Conclusion**
7. **Test Your Conclusion**

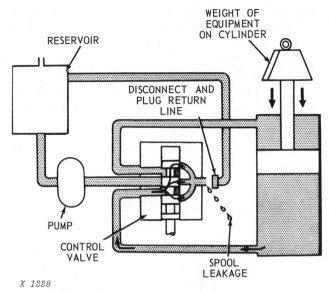

X 1228

Fig. 9—Checking Control Valve for Leaks

CHECKING FOR LEAKS

If you suspect the control valve or cylinder is leaking, do the following:

Raise the hydraulic equipment a few feet off the ground, return the control lever to neutral and shut off the engine.

Notice whether the equipment settles toward the ground. If the equipment settles, temporarily support it and disconnect the return line between the control valve and reservoir, then plug the line (Fig. 9).

Remove the support and examine the open port in the control valve as the equipment settles. If oil leaks from the port, the control valve spool is leaking.

If no oil is leaking from the control valve, check the cylinder as follows:

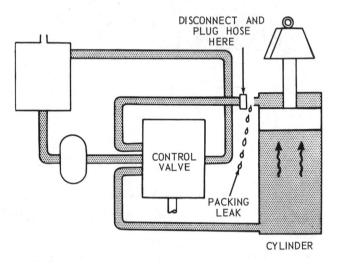

X 1229

Fig. 10—Checking Double-Acting Cylinder for Leaks

To check a double-acting cylinder, run the cylinder to one end of its stroke. Support the equipment if it is raised, then shut off the engine. Remove the hose from the end of the cylinder that was not pressurized (Fig. 10). Start the engine again, pressurize the cylinder, and see if any oil comes out of the open port. Repeat the test in the opposite direction since it may be possible for the cylinder to leak in only one direction. If oil leaks out the open cylinder port, the packings in the cylinder must be replaced.

TESTING THE MACHINE

Testing with gauges or a hydraulic analyzer is the most effective way to pinpoint troubles in the system, but it may not be the most efficient way.

However, here are some preliminary checks you can make without using a tester or prior to using one.

OPERATIONAL CHECKOUT

Some companies have designed a system of diagnosing machine performance called "Operational Checkout." This diagnostic approach is helpful in pinpointing many machine malfunctions without the use of gauges, sensors or other diagnostic tools. Operational checkout is also referred to as the Look — Listen — Feel approach.

HYDRAULIC SYSTEM CHECKS

Look-Listen-Feel

The operational checkout procedures must be followed step-by-step to be effective and efficient. The following operational checkout will give you an idea how simply some of the hydraulic system malfunctions can be detected and isolated. So, look, listen and feel your way to more rapid trouble shooting.

| **HYDRAULIC PUMP PERFORMANCE CHECK** |  | *NOTE: If hydraulic oil is not at operating temperature, heat hydraulic oil until loader and backhoe feel warm to touch using following procedure:*

Put backhoe in transport position and engage boom and swing lock.

Activate boom down function and run engine at 2000 rpm.

NOTE: If activating boom down does not load engine, boom down relief valve is set too high or hydraulic pump stand-by pressure is too low.

Operate all functions periodically to distribute heated oil to all cylinders.

Continued on next page |

Put backhoe at maximum reach with bucket fully dumped at ground level.

Run engine at 2000 rpm.

Measure cycle time by simulating loading the bucket, retracting the dipperstick and raising the boom to the boom cylinder cushion. Do not time boom cylinder through cushion.

LOOK: The maximum cycle time is as follows:

410D - 9 seconds

510D - 12 seconds

NOTE: Take the average cycle time for at least 3 complete cycles. This average cycle time will give a general indication of hydraulic pump performance.

OK: Go to next check.

NOT OK: Replace hydraulic filter and check for types of contamination. Rerun this check.

NOT OK: Cycle times still slow. Go to Backhoe Relief Valve Test.

BACKHOE CIRCUIT LEAKAGE CHECK

Put backhoe in transport position and engage boom lock.

Retract extendible dipperstick (if equipped).

Raise stabilizers to full up position.

Run engine at slow idle.

Fully activate functions, one at a time:
Boom up
Bucket load
Dipperstick retract
Extendible dipperstick (if equipped).

LISTEN: When these functions are activated, NO decrease in engine rpm must be noted.

Fully activate functions, one at a time:
Boom down.
Swing left then right.
Raise both stabilizers.

LISTEN: Boom down must cause rpm to decrease since relief valve setting is below standby pressure.

LISTEN: Swing left and right may cause rpm to decrease slightly because relief valve setting is close to standby pressure.

LISTEN: Stabilizer circuit can cause rpm to decrease slightly because of normal valve leakage.

OK: Go to next check.

NOTE: Pressure seal passages are used in boom valve and leakage return passages are used in swing valve. When boom down or swing functions are bottomed and control valves "metered", rpm will decrease and leakage within circuit will be apparent. This is normal.

NOT OK: Continue on.

BACKHOE CIRCUIT LEAKAGE CHECK—CONTINUED

Lower stabilizer to maximum down position, extend dipperstick to maximum reach, extend extendible dipperstick (if equipped) and put bucket in dump position 1 m (3 ft) off ground.

Fully activate functions, one at a time:
Extend dipperstick
Extend extendible dipperstick (if equipped)
Bucket dump

Fully activate functions, one at a time:
Stabilizer down left
Stabilizer down right

LISTEN: When these functions are activated, NO decrease in engine rpm must be noted.

OK: Go to next check.

NOT OK: If rpm decreases with a function bottomed and control lever fully open, a leak is indicated in the circuit.

NOT OK: If rpm increases when a function is bottomed and control valve is fully open, a neutral leak indicated.

NOT OK: A rpm decrease with a function bottomed in both directions is normally cylinder leakage. A rpm decrease in one direction is normally circuit relief valve leakage. Go to Hydraulic Component Leakage Test.

LOADER CIRCUIT LEAKAGE CHECK

Raise loader to full height and put bucket in dump position.

Run engine at slow idle.

Fully activate functions, one at a time:
Loader boom up
Bucket dump

Put bucket in rollback position and lower loader to full down position.

Fully activate functions, one at a time:
Loader boom down
Bucket rollback

LISTEN: When functions are activated, engine rpm must NOT decrease.

NOTE: Leakage return passages are used in boom valve. When boom cylinders are retracted and valve is "metered", rpm will decrease and leakage within the circuit will be apparent. This is normal.

OK: Go to next check.

NOT OK: If rpm decreases with a function bottomed and control lever is fully open, a neutral leak indicated.

NOT OK: A rpm decrease with a function bottomed in both directions is normally cylinder leakage. A rpm decrease in one direction is normally circuit relief valve leakage. Go to Hydraulic Component Leakage Test.

CYLINDER CUSHION CHECK	Run engine at approximately 1000 rpm. Activate backhoe swing left and right and boom raise. Note sound and speed as cylinders near end of their stroke. *LOOK: Speed of cylinder rod must decrease near the end of its stroke.*	*LISTEN: Must hear oil flowing through orifice as cylinder rod nears the end of its stroke.*	**OK:** Go to next check. **NOT OK:** Remove and repair cylinder cushion. Go to repair manual.

BACKHOE AND LOADER FUNCTION DRIFT CHECK	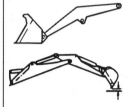	*FEEL: Backhoe cylinders. Cylinders must be warm to touch 38 — 52°C (100—125°F). If cylinders are not warm, heat hydraulic oil to specification, (see Group 9025-25).* Raise unit off ground with stabilizers. Put backhoe bucket at a 45° angle to ground. Lower boom until bucket cutting edge is 50 mm (2 in.) off ground. Position loader bucket same angle and distance off ground as backhoe bucket. Run engine at slow idle and observe buckets cutting edges. *LOOK: If bucket cutting edges touch the ground within 1 minute, leakage is indicated in the bucket or boom cylinders or control valves.*	**OK:** Go to next check. **NOT OK:** Use good judgment in determining if the amount of drift is objectionable for the type of operation the unit is performing. **NOT OK:** Isolate which function is leaking. Go to Cylinder Drift Test.

LOADER BOOM FLOAT AND RETURN-TO-DIG CHECK	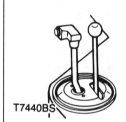 T7440BS	Put loader at maximum height position with bucket dumped. Run engine at approximately 2000 rpm. Move the loader control lever forward into boom float detent position, and at the same time into bucket rollback detent position. Remove hand from control lever. *LOOK: Loader control lever must remain in the boom float detent position.*	**OK:** Continue On. **NOT OK:** If lever jumps out of detent, inspect detent spring and balls. Go to repair manual.

Many malfunctions are quickly identified with these **Operational Checkout** procedures, but **Look-Listen-Feel** can only go so far in a complex system. If the problem needs further investigation, it is time to use hydraulic testing equipment and procedures as you will learn on the following pages.

WHAT A HYDRAULIC TESTER DOES

If you cannot locate the trouble with the operational checks, check the system using a hydraulic analyzer or test gauges. Using this equipment, you can accurately measure oil flow, pressure, and temperature, and quickly isolate faulty components.

Hydraulic analyzers are available with pressure loading valves, pressure gauges (high and low pressure), flow meters, and temperature gauges to precisely analyze a complex hydraulic system.

In testing any hydraulic circuit, the following four checks are of prime importance:

1. *Temperature*—The oil should be checked for correct operating temperature to assure accuracy of the tests that follow.

2. *Flow*—The flow check determines if the pump is developing its rated output.

3. *Pressure*—Pressure checks test relief valves for proper operation. (In a closed-center system, pressure checks indicate the operation of the main pump.)

4. *Leakage*—The leakage test isolates leakage in a particular component.

These basic checks may be made with most hydraulic testers. Before beginning, however, read the instruction manual furnished with the tester and review the system. You should have a thorough knowledge of the machine's specifications (relief valve pressures, pump output, engine rpm, and operating temperature) to accurately test the system. Refer to the machine's repair manual.

To test a machine, you must disconnect some of the oil lines. But remember, DIRT IS THE WORST ENEMY OF A HYDRAULIC SYSTEM. Before disconnecting oil lines, steam clean the machine. And be sure to plug all openings to keep out dirt.

PUMP TESTING

The pump is the generating force for the whole hydraulic system. This is the place to start testing the system.

Installing the Hydraulic Tester

1. Relieve any pressure in the system and disconnect the pressure line between the pump and the control valve. Attach the pressure line to the hydraulic tester INLET port (Fig. 11).

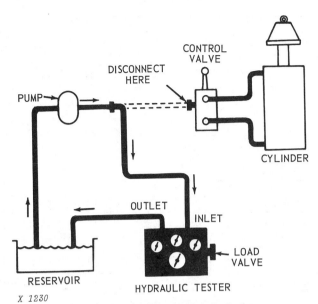

X 1230

Fig. 11—Testing the Hydraulic Pump

2. Connect hydraulic tester OUTLET port to the reservoir. Whenever possible, connect directly to the reservoir return line because it usually has a return filter. On a closed-center system, always return hydraulic tester oil to a point between the main hydraulic pump and the charging pump to maintain pressure in the system (or to be sure the main pump does not lose its charge).

3. Check the oil level and slowly close the tester load valve to load the system. (Do not exceed the system's maximum rated pressure.) Continue loading until the normal operating temperature of the system is reached (see machine specifications).

Be sure the tester load valve is OPEN before starting any tests. The load valve can develop tremendous pressure on a component if it is closed too far.

Operating the Hydraulic Tester

1. With the tester load valve open, record maximum pump flow at zero pressure.

2. Slowly close the load valve to increase pressure and record the flow at 250 psi increments from zero pressure to maximum system pressure. Write down your test results so you can refer to them later. Use a test form such as the one shown in Fig. 12 on the next page.

3. Open the hydraulic tester load valve until maximum pump flow is again at zero pressure.

4. Shut off the engine.

			Gallons Per Minute @ Psi								
		0	250	500	750	1000	1250	1500	1750	2000	2250
Pump Test		31.0	28.0	25.0	22.0	19.0	16.0	13.0	10.0	7.0	
Circuit Test	Direction Cylinder Travel										

X 1231

Fig. 12—Sample Form for Recording Pump Test Results

Pump Test Diagnosis

Pump flow at maximum pressure should be at least 75 percent of pump flow at zero pressure. (On modern variable displacement pumps of the radial piston type, 90 percent can be expected.) A lower reading like the one shown on the form in Fig. 12 indicates a badly worn pump.

If pump flow is poor during the free flow test as well as the pressure tests, the pump probably is not getting enough oil. This problem could be caused by low oil supply, air leaks, a restricted pump inlet line, or a dirty reservoir, filter or breather.

If the pump tests okay, then start checking the system components for trouble.

SYSTEM TESTING

Installing the Hydraulic Tester

Install a tee fitting in the line between the pump and the control valve and attach the hydraulic tester INLET port to this tee (Fig. 13).

Leave the return line from the hydraulic tester OUTLET port connected in the same way as it was for the pump test.

Operating the Hydraulic Tester

1. Open the hydraulic tester load valve.

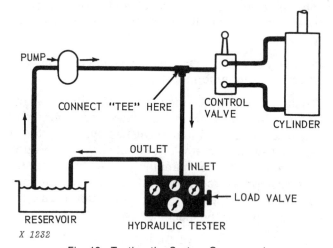

X 1232

Fig. 13—Testing the System Components

2. Start the engine and adjust it to the manufacturer's recommended operating speed.

3. Slowly close the hydraulic tester load valve to load the system. Continue loading the system until normal operating temperature is reached.

4. Open load valve to record maximum system flow at zero pressure.

5. Operate the control valve and hold it in one of its power positions.

		Gallons Per Minute @ Psi									
		0	250	500	750	1000	1250	1500	1750	2000	2250
Pump Test		32.0	31.7	31.4	31.0	30.6	30.2	29.8	29.3	28.8	
Circuit Test	Direction Cylinder Travel										
Boom Circuit	Lower	32.0	31.7	31.4	31.0	30.6	30.2	29.8	29.3	28.8	
	Raise	32.0	31.7	31.4	31.0	30.6	30.2	29.8	29.3	28.8	
Bucket Circuit	Dump	32.0	31.7	31.4	31.0	5.0	0.0				
	Roll-Bk.	32.0	31.7	31.4	31.0	30.6	30.2	10.0	0.0		

X 1233

Fig. 14—Sample Form for Recording System Test Results

6. Slowly close the hydraulic tester load valve and record flow in 250 psi increments from zero pressure to maximum system pressure (Fig. 14).

7. Open the load valve until maximum flow is again at zero pressure and repeat the test in the rest of the control valve power positions.

Be sure to make all the tests at the same oil temperature to get readings that can be compared. If oil is too hot from the previous test, allow it to circulate through the system for cooling.

System Test Diagnosis

Here's how to judge the system tests:

1. *If flow at each pressure is same as for pump test:* All components are okay.

2. *If pressure begins to drop before full load is reached:* One of the circuits is bad. (Such as the bucket roll-back circuit in Fig. 14).

The pressure drop is caused by leakage. To find out whether the leakage is in the control valve or the cylinder, disconnect the cylinder return line and move the control valve to a power position. If oil leaks from the cylinder return port, the cylinder is at fault and must be repaired. If no oil leaks out, the control valve is probably at fault.

3. *If flow drops the same with the control valve in all positions:* The system relief valve is probably at fault (see following). This condition could also indicate a leak in the control valve.

Relief Valve Diagnosis

If equipment with circuit relief valves is being checked, you can tell when the valves open because flow will suddenly drop about 3 gpm (or drop to zero gpm if a full-flow relief valve is used). Often the relief valves will start to "crack" open before they reach their full-pressure settings. This can be noted by comparing the pressure and flow rate readings made in the circuit test above. Any great decrease in flow rate in these valves indicates a faulty valve.

As a general rule:

Faulty SYSTEM relief valves will affect readings in all tests.

Faulty CIRCUIT relief valves will affect only pressure readings in the individual circuits.

OPTIONAL TEST EQUIPMENT

The hydraulic tester used in the preceding discussion contains a flow meter, pressure gauges, and temperature gauges. For more complex systems, each piece of test equipment may be used separately.

Flow Meter

The flow meter in Fig. 15 is used to check hydraulic oil flow and pressure.

Pressure Gauges

The pressure gauges and connectors in Fig. 16 are used to check hydraulic pressures.

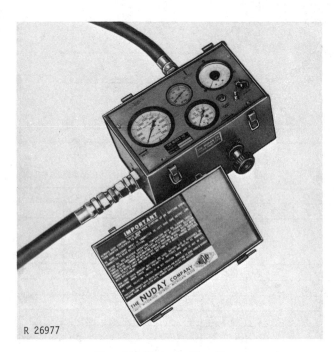

R 26977

Fig. 15 — Hydraulic Flow Meter

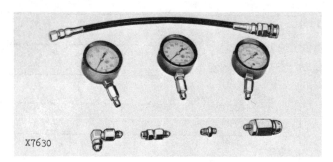

X7630

Fig. 16 — Pressure Gauges

SUMMARY: TESTING THE MACHINE

The tests we have given you are only basic guidelines. Once you start testing actual machines, use your machine technical manual for detailed tests and test results. And remember that the best testing equipment has no value unless the man at the controls knows how to interpret the results.

TROUBLE SHOOTING CHARTS

INTRODUCTION

Use the charts on the following pages to help in listing all the possible causes of trouble when you begin diagnosis and testing of a machine.

Before starting any testing, first check for external oil leaks, return lines and passages for excessive heat due to internal oil leaks, and unusual noises in the system.

Once you have located the cause, check the item in the chart again for the possible remedy.

The technical manual for each machine supplements these charts by giving more detailed and specific causes and remedies.

HYDRAULIC OIL CONDITION

Oil Milky or Dirty

Water in oil (milky).

Filter failures (dirty).

Metal particles (mechanical failure).

Oil Discolored or Has Burned Odor

Kinked pipes.

Plugged oil cooler.

Wrong oil viscosity.

Internal leaks.

SYSTEM INOPERATIVE

No oil in system.

Fill to full mark. Check system for leaks.

Oil low in reservoir.

Check level and fill to full mark. Check system for leaks.

Oil of wrong viscosity.

Refer to specifications for proper viscosity.

Filter dirty or plugged.

Drain oil and replace filters. Try to find source of contamination.

Restriction in system.

Oil lines could be dirty or have inner walls that are collapsing to cut off oil supply. Clean or replace lines. Clean orifices.

Air leaks in pump suction line.

Repair or replace lines.

Dirt in pump.

Clean and repair pump. If necessary, drain and flush hydraulic system. Try to find source of contamination.

Badly worn pump.

Repair or replace pump. Check for problems causing pump wear such as misalignment or contaminated oil.

Badly worn components.

Examine and test valves, motors, cylinders, etc. for external and internal leaks. If wear is abnormal, try to locate the cause.

Oil leak in pressure lines.

Tighten fittings or replace defective lines. Examine mating surfaces on couplers for irregularities.

Components not properly adjusted.

Refer to machine technical manual for proper adjustment of components.

Relief valve defective.

Test relief valves to make sure they are opening at their rated pressure. Examine seals for damage that could cause leaks. Clean relief valves and check for broken springs, etc.

Pump rotating in wrong direction.

Reverse to prevent damage.

Operating system under excessive load.

Check specifications of unit for load limits.

Hoses attached improperly.

Attach properly and tighten securely.

Slipping or broken pump drive.

Replace couplers or belts if necessary. Align them and adjust tension.

Pump not operating.

Check for shut-off device on pump or pump drive.

SYSTEM OPERATES ERRATICALLY

Air in system.

Examine suction side of system for leaks. Make sure oil level is correct. (Oil leak on the pressure side of the system could account for loss of oil.)

Cold oil.

Viscosity of oil may be too high at start of warm-up period. Allow oil to warm up to operating temperature before using hydraulic functions.

Components sticking or binding.

Check for dirt or gummy deposits. If dirt is caused by contamination, try to find the source. Check for worn or bent parts.

Pump damaged.

Check for broken or worn parts. Determine cause of pump damage.

Dirt in relief valves.

Clean relief valves.

Restriction in filter or suction line.

Suction line could be dirty or have inner walls that are collapsing to cut off oil supply. Clean or replace suction line. Also, check filter line for restrictions.

SYSTEM OPERATES SLOWLY

Cold oil.

Allow oil to warm up before operating machine.

Oil viscosity too heavy.

Use oil recommended by the manufacturer.

Insufficient engine speed.

Refer to operator's manual for recommended speed. If machine has a governor, it may need adjustment.

Low oil supply.

Check reservoir and add oil if necessary. Check system for leaks that could cause loss of oil.

Adjustable orifice restricted too much.

Back out orifice and adjust it. Check machine specifications for proper setting.

Air in system.

Check suction side of the system for leaks.

Badly worn pump.

Repair or replace pump. Check for problems causing pump wear such as misalignment or contaminated oil.

Restriction in suction line or filter.

Suction line could be dirty or have inner walls that are collapsing to cut off oil supply. Clean or replace suction line. Examine filter for plugging.

Relief valves not properly set or leaking.

Test relief valves to make sure they are opening at their rated pressure. Examine valves for damaged seats that could leak.

Badly worn components.

Examine and test valves, motors, cylinders, etc. for external and internal leaks. If wear is abnormal, try to locate the cause.

Valve or regulators plugged.

Clean dirt from components. Clean orifices. Check source of dirt and correct.

Oil leak in pressure lines.

Tighten fittings or replace defective lines. Examine mating surfaces on couplers for irregularities.

Components not properly adjusted.

Refer to machine technical manual for proper adjustment of components.

SYSTEM OPERATES TOO FAST

Adjustable orifice installed backward or not installed.

Install orifice parts correctly and adjust.

Obstruction or dirt under seat of orifice.

Remove foreign material. Readjust orifice.

OVERHEATING OF OIL IN SYSTEM

Operator holds control valves in power position too long, causing relief valve to open.

Return control lever to neutral position when not in use.

OVERHEATING OF OIL IN SYSTEM—Continued

Using incorrect oil.

Use oil recommended by manufacturer. Be sure oil viscosity is correct.

Low oil level.

Fill reservoir. Look for leaks.

Dirty oil.

Drain and refill with clean oil. Look for source of contamination.

Engine running too fast.

Reset governor or reduce throttle.

Incorrect relief valve pressure.

Check pressure and clean or replace relief valves.

Internal component oil leakage.

Examine and test valves, cylinders, motors, etc. for external and internal leaks. If wear is abnormal, try to locate cause.

Restriction in pump suction line.

Clean or replace.

Dented, obstructed or undersized oil lines.

Replace defective or undersized oil lines. Remove obstructions.

Oil cooler malfunctioning.

Clean or repair.

Control valve stuck in partially or full open position.

Free all spools so that they return to neutral position.

Heat not radiating properly.

Clean dirt and mud from reservoir, oil lines, coolers, and other components.

Automatic unloading control inoperative (if equipped).

Repair valve.

FOAMING OF OIL IN SYSTEM

Low oil level.

Fill reservoir. Look for leaks.

Water in oil.

Drain and replace oil.

Wrong kind of oil being used.

Use oil recommended by manufacturer.

Air leak in line from reservoir to pump.

Tighten or replace suction line.

Kink or dent in oil lines (restricts oil flow).

Replace oil lines.

Worn seal around pump shaft.

Clean sealing area and replace seal. Check oil for contamination or pump for misalignment.

PUMP MAKES NOISE

Low oil level.

Fill reservoir. Check system for leaks.

Oil viscosity too high.

Change to lighter oil.

Pump speed too fast.

Operate pump at recommended speed.

Suction line plugged or pinched.

Clean or replace line between reservoir and pump.

Sludge and dirt in pump.

Disassemble and inspect pump and lines. Clean hydraulic system. Determine cause of dirt.

Reservoir air vent plugged.

Remove breather cap, flush, and clean air vent.

Air in oil.

Tighten or replace suction line. Check system for leaks. Replace pump shaft seal.

PUMP MAKES NOISE—Continued

Worn or scored pump bearings or shafts.

Replace worn parts or complete pump if parts are badly worn or scored. Determine cause of scoring.

Inlet screen plugged.

Clean screen.

Broken or damaged pump parts.

Repair pump. Look for cause of damage like contamination or too much pressure.

Sticking or binding parts.

Repair binding parts. Clean parts and change oil if necessary.

PUMP LEAKS OIL

Damaged seal around drive shaft.

Tighten packing or replace seal. Trouble may be caused by contaminated oil. Check oil for abrasives and clean entire hydraulic system. Try to locate source of contamination. Check the pump drive shaft. Misalignment could cause the seal to wear. If shaft is not aligned, check the pump for other damage.

Loose or broken pump parts.

Make sure all bolts and fittings are tight. Check gaskets. Examine pump castings for cracks. If pump is cracked, look for a cause like too much pressure or hoses that are attached incorrectly.

LOAD DROPS WITH CONTROL VALVE IN NEUTRAL POSITION

Leaking or broken oil lines from control valve to cylinder.

Check for leaks. Tighten or replace lines. Examine mating surfaces on couplers for irregularities.

Oil leaking past cylinder packings or O-rings.

Replace worn parts. If wear is caused by contamination, clean hydraulic system and determine the source.

Oil leaking past control valve or relief valves.

Clean or replace valves. Wear may be caused by contamination. Clean system and determine source of contamination.

Control lever not centering when released.

Check linkage for binding. Make sure valve is properly adjusted and has no broken or binding parts.

CONTROL VALVE STICKS OR WORKS HARD

Misalignment or seizing of control linkage.

Correct misalignment. Lubricate linkage joints.

Tie bolts too tight (on valve stacks).

Use manufacturer's recommendation to adjust tie bolt torque.

Valve broken or scored internally.

Repair broken or scored parts. Locate source of contamination that caused scoring.

CONTROL VALVE LEAKS OIL

Tie bolts too loose (on valve stacks).

Use manufacturer's recommendation to adjust tie bolt torque.

Worn or damaged O-rings.

Replace O-rings (especially between valve stacks). If contamination has caused O-rings to wear, clean system and look for source of contamination.

Broken valve parts.

If valve is cracked, look for a cause like too much pressure or hoses that are attached incorrectly.

CYLINDERS LEAK OIL

Damaged cylinder barrel.

Replace cylinder barrel. Correct cause of barrel damage.

Rod seal leaking.

Replace seal. If contamination has caused seal to wear, look for source. Wear may be caused by external as well as internal contaminants. Check piston rod for scratches or misalignment.

Loose parts.

Tighten parts until leakage has stopped.

Piston rod damaged.

Check rod for nicks or scratches that could cause seal damage or allow oil leakage. Replace defective rods.

CYLINDER LOWERS WHEN CONTROL VALVE IS IN "SLOW RAISE" POSITION

Damaged check valve in lift circuit.

Repair or replace check valve.

Leaking cylinder packing.

Replace packing. Check oil for contamination that could cause wear. Check alignment of cylinder.

Leaking lines or fittings to cylinder.

Check and tighten. Examine mating surfaces on couplers for irregularities.

POWER STEERING DOES NOT WORK, STEERS HARD, OR IS SLOW

Air in system.

Bleed system. Check for air leaks.

Internal leakage in system.

Components may not be adjusted properly. Parts may be worn or broken. Check for cause of wear.

System not properly timed.

Time according to manufacturer's instructions.

Worn or damaged bearings.

Check and replace bearings in steering components.

Insufficient pressure.

Check pump and relief valves. Contamination could cause valves to leak or pump to wear.

POWER BRAKES MALFUNCTION

Heavy oil or improper brake fluid.

Warm up fluid or change to one of lighter viscosity. Use proper oil or brake fluid (see machine operator's manual).

NOTE: Many brake circuits use brake fluid instead of hydraulic oil. DO NOT MIX.

Air in system.

Bleed brake system. Find out where air is coming from.

Contaminated oil.

This may cause components to wear or jam. Clean and repair system and check for cause of contamination.

Brake pedal return restricted.

Clean dirt from moving parts. Check linkage for damage.

Accumulator not working (if equipped).

Check accumulator precharge. If accumulator is defective, repair or replace it.

TEST YOURSELF
QUESTIONS

1. Give the seven basic steps for good trouble shooting.

2. During which of the seven steps should you begin replacing parts?

3. True or false? "Test the system flow first so that you have a guide for readings on the pump flow tests."

4. (Fill in the blanks with "circuit" or "system".) "As a general rule, faulty _____ relief valves will affect pressure readings on all tests, while faulty _____ relief valves will affect only some readings."

5. What is an "Operational Checkout"?

6. What three senses are used during an Operational Checkout?

SYMBOLS USED IN FLUID POWER DIAGRAMS/CHAPTER 13

READING HYDRAULIC SYMBOL DRAWINGS

Symbolic drawings and diagrams are the most popular way of representing hydraulic components and systems. These symbols are easier to draw, read, and standardize than the other types of engineering drawings. In order to be literate in the maintenance and manufacturing process, you must understand the rules that make symbolic drawing such a unique language.

In this chapter we will look at the guidelines that you can use to interpret and understand symbolic drawings. We will also look at how components are interconnected to make a working hydraulic system.

As opposed to pictorial drawings, symbols do not show the exact shape of the component that they represent. Symbols are two dimensional. They are lines drawn on paper. They have no additional explanation (such as color coding or words) (Fig. 1).

Instead symbols are a visual short-hand method of communication. They rely on figures, such as squares and circles, and marks, such as arrows, to represent hydraulic components. Unlike cutaway drawings, symbols do not show the parts of the hydraulic component. Symbols and symbolic diagrams do show:

RELIEF VALVE

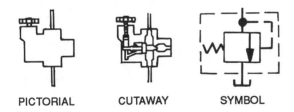

PICTORIAL   CUTAWAY   SYMBOL

Fig. 1 — Symbols Do Not Show the Visual Detail of a Pictorial or Cutaway Drawing.

• How components are interconnected in a system (Fig. 2).

• Flow paths of hydraulic fluid.

• The general way a component works.

• The number of ports or connections on the component.

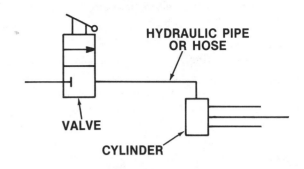

Fig. 2 — Symbolic Diagrams Show How Components are Interconnected.

Some basic guidelines that you can use in this chapter to help interpret symbols are:

• Symbols and diagrams do not show the internal conditions of the hydraulic system. This includes fluid temperature and pressure.

• Each symbol is drawn to represent the neutral or normal position of the component before it is actuated. A normally closed value will be shown in the closed position (Fig. 3).

• Symbols can be rotated or put into a location that is not their normal position. This does not alter their meaning. They must be shown correctly connected if they are drawn in a hydraulic system diagram.

• Symbols can be drawn in any size without altering their meaning.

• All line width should be read as the same thickness. Unlike orthographic drawing that relies on line width to convey an idea, symbols are straight forward and do not need to be interpreted.

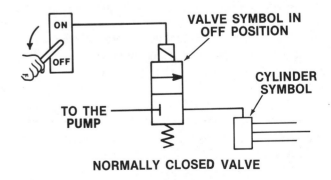

NORMALLY CLOSED VALVE

Fig. 3 — The Normal or Nonactuated State of the Component is Represented by the Symbol.

CHARACTERISTICS OF SYMBOLS

The graphical symbols used as examples in this chapter are in accordance with ISO 1219-1 (Fluid Power Systems and Components — Graphical Symbols and Circuit Diagrams — Part 1: Graphical Symbols).

The symbols that we look at are **based** on this ISO standard. Any **color**, words, or lines outside of the symbol are not part of the symbol. These external marks have been added to help clarify some particular point.

SHAPES

The basic shapes used to make hydraulic symbols are:

- **Circles, semicircles**
- **Squares**
- **Diamonds**
- **Rectangles**

MARKS

A symbol is made by using one of the four basic shapes and adding the appropriate marks. These marks include:

- **Lines**
- **Arrows, arrowheads**
- **Arcs**

Circles, Semicircles

When you see a circle or a semicircle with marks inside of it, you are looking at a pump or motor symbol. The circle represents circular rotation such as the internal rotating parts of a motor (Fig. 4).

Squares

Squares are also called **envelopes**. They are used to represent one position or hydraulic fluid path through a valve. Two envelopes together make a two **position** valve. Three together make a three position valve. Four together make a four position valve.

Diamonds

Diamonds represent a part that helps condition the hydraulic fluid in a system. These include fluid coolers, and filters.

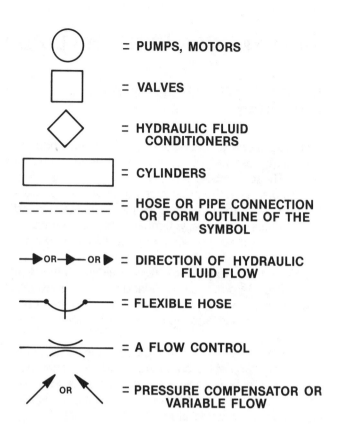

Fig. 4 — There are Four Basic Shapes, a Broken or Unbroken Line, and Five Basic Marks That are Used to Make Hydraulic Symbols.

Rectangles

Rectangles represent hydraulic cylinders.

Lines

There are four basic types of lines (Fig. 5).

- The solid line represents the route that hydraulic fluid flows through the system.

- The dash shows a pilot line that connects from a pressurized part of the system back to a part that helps control the system. This helps keep the system stable.

- Center lines are used to enclose assemblies.

- The double line is used to show a mechanical connection (shaft, rod, lever, etc.).

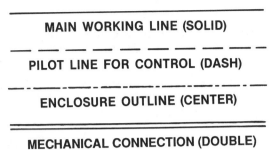

MAIN WORKING LINE (SOLID)

PILOT LINE FOR CONTROL (DASH)

ENCLOSURE OUTLINE (CENTER)

MECHANICAL CONNECTION (DOUBLE)

Fig. 5 — There Are Four Symbols For Lines.

Lines (pipes or hose) that cross in a hydraulic system, but do not connect, are represented as uninterrupted lines. They are also shown as arcs looped over one another (Fig. 6). Lines that show hose or tubing that connect are joined by a bullet, or are shown as perpendicular joining lines (Fig. 7).

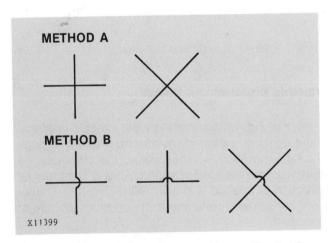

METHOD A

METHOD B

X11399

Fig. 6 — Uninterrupted Lines Represent Hose or Pipe that Cross Over Each Other But Do Not Connect.

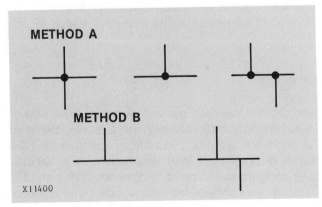

METHOD A

METHOD B

X11400

Fig. 7 — Connected Pipes or Hose Symbols

Arrows

Arrows show the direction of the flow of hydraulic fluid. They also show direction of movement in pumps and motors. An arrow inside of a circle or through two arcs can represent an adjustment point for the amount of hydraulic fluid flow or pressure (Fig. 8).

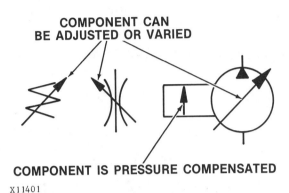

COMPONENT CAN
BE ADJUSTED OR VARIED

COMPONENT IS PRESSURE COMPENSATED

X11401

Fig. 8 — Arrows Can Represent Adjustment Points.

Arcs

Arcs show a point of adjustment when they are used together. See Fig. 4 "a flow control." They are also used to show a flexible hose line in a hydraulic system.

HYDRAULIC COMPONENTS AND SYMBOLS

PUMPS

The basic symbol for hydraulic pumps is a circle or semicircle (Fig. 9). The circle or semicircle alone isn't enough. It doesn't tell you anything at all about the pump. For example it doesn't tell you how many ports it has, whether it has fixed or variable displacement; whether or not it is unidirectional or bidirectional (sometimes referred to as over-center); or if it is pressure compensated.

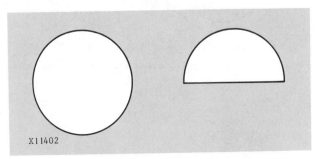

X11402

Fig. 9 — Circles and Semicircles Represent Pumps and Motors.

Even the cutaway and pictorial views shown in Fig. 10 do not tell us much more than the ISO symbol. We **can** tell from the cutaway and pictorial views that the pump shown is an external gear pump (Chapter 2). We can also see how the pump is constructed and how the gears, shafts, wear plates, and housing fit together.

When a hydraulic system is shown, the ISO symbol tells us all we need to know. It represents the function of the pump within the circuit and its relationship to the other components that make up the circuit.

Now we will start adding some arrows, lines, arrow-heads, and rectangles to it and see how we can use a circle to represent many different types of pumps.

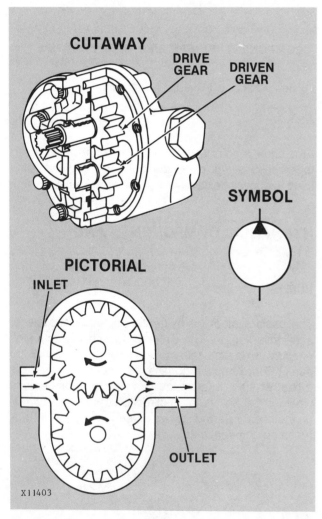

Fig. 10 — Symbols Often Tell You As Much Or More Information Than Cutaway Or Pictorial Drawings.

Fixed Displacement Pumps

When we look at Fig. 11, we start to learn a little more. View 1 symbolizes a unidirectional pump with two parts. View 2 is a bidirectional pump (hydraulic oil can flow in either direction) as indicated by the two triangles. Both pumps shown in these two views are fixed displacement pumps.

Now, let's add some arrows and see how they affect our symbol.

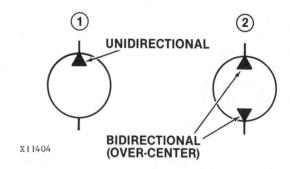

Fig. 11 — Arrows Show Oil Flow in Pumps.

Variable Displacement Pumps

View 1 of Fig. 12 is a unidirectional variable displace-ment pump as indicated by the arrows going through the circle symbol. In other words, the output of the pump can be adjusted or varied. Also, a shaft and its direction of rotation is shown. When indicating direc-tion, the arrow is assumed to be on the near side of the shaft.

View 2 of Fig. 12 is an even more sophisticated pump. It is a bidirectional, variable displacement, pressure compensated pump with a reversing shaft and a drain line back to the hydraulic reservoir of the machine. Notice the arrowhead at the bottom of the circle.

Summary

The circle is the basic symbol for pumps. The type of pump the circle is representing will determine the need for additional symbols. The addition of arrows tell if the pump is pressure compensated, if it is a variable displacement type, and in what direction the shaft is rotating. Arrowheads indicate whether it is unidirec-tional or bidirectional.

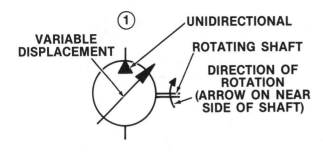

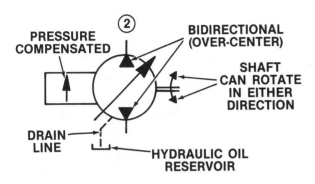

Fig. 12 — Complex Pump Symbols Are Made From Basic Symbols.

MOTORS

The basic symbol for a hydraulic motor is the same as for a pump. Motors and pumps are similar in design (see Chapter 5).

Like pumps, motors can be either unidirectional or bidirectional. They can be fixed or variable displacement, and be pressure compensated. These characteristics can all be shown symbolically.

It is not enough to show a circle with the parts added. You would have no way of knowing whether the symbol represents a pump or motor. It is represented by simply inverting the arrowheads within the circle symbol. This also indicates whether the pump or motor is unidirectional or bidirectional. Since all pumps and motors are either unidirectional or bidirectional, arrowheads indicate the type of pump or motor.

In Fig. 13 we see a bidirectional, variable displacement, pressure compensated pump and motor with a reversing shaft and drain line. View 1 is the pump; view 2 is the motor.

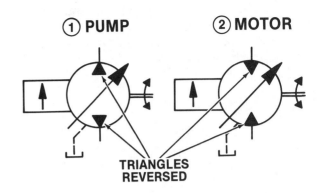

Fig. 13 — The Position of Arrowheads Differentiate Pumps from Motors.

Both Pump and Motor

In some hydraulic systems, a component will operate as a pump part of the time and as a motor part of the time, depending on what the system was designed to do.

How do we read this on our fluid power diagrams? Again, it's simply the way the arrowheads are placed within the circle symbol.

Let's look at some examples.

In Fig. 14 we see the symbol for a part that is both pump and motor. View 1 indicates that the component operates in one direction as a pump and in the other direction as a motor. View 2 shows a component that operates in only one direction but either as a pump or motor. Note the placement of the arrowheads.

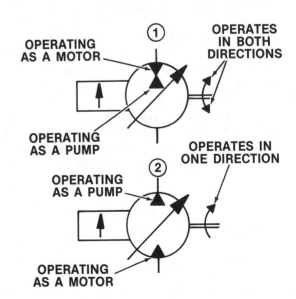

Fig. 14 — Pumps Can Operate as Motors.

The component shown in Fig. 15 can operate as a pump or motor in either direction. Note the four arrowheads. Two indicate a bidirectional pump and two indicate a bidirectional motor.

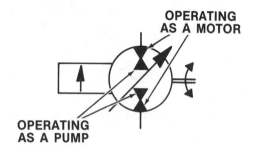

Fig. 15 — A Bidirectional Pump That Can Operate as a Motor Has Two Sets of Arrowheads.

Summary

The circle is the basic symbol for a motor and a pump. Arrows and arrowheads indicate the same things for a motor that they do for a pump. The key to whether a circle is representing a motor or a pump is the way that the arrowheads are placed. This also indicates a unidirectional or bidirectional working of the motor or pump. In a pump symbol, the arrowheads point toward the outer circumference of the circle; in a motor symbol, they point toward the center of the circle.

VALVES

As discussed in Chapter 3, there are three major types of valves. They are the directional control, pressure control, and volume control valve. Three different types of ANSI symbols are used to represent them.

The square is the basic ISO symbol used to represent one of the most common directional valves — the spool valve (Fig. 16). Two or more squares or envelopes indicate a valve having as many distinct positions as there are squares. Before we go farther in our discussion of spool valves, let us review some rules in the correct use of ISO symbols for spool valves.

Correct Usage of ISO Symbols in Directional Control Spool Valves

ISO symbols are internationally used and recognized. Because of this, certain rules must be followed when using them in fluid power diagrams to represent valves.

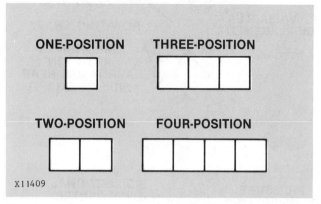

Fig. 16 — Squares Symbolize One Possibility of Oil FLow in a Valve.

1. Since the valve can only be in one valving position at a time, all connections from the valve to the rest of the circuit must be made from one block or envelope only. See Fig. 17 "hydraulic line to rest of circuit."

2. All connections to the external circuit from the ISO symbol should be made from the square which shows flow when the valve is in its nonactuated or normal state. In Fig. 17 we see a 2-way normally closed spool valve with a manual actuating lever and return spring. In its nonactuated or normal state, the valve is closed. Therefore, it is shown connected to the rest of the circuit from the square that indicates no flow or closed position.

3. All valve actuators — hand lever, foot pedal, and solenoid — should always be visualized as "pushing" the entire assembly of porting squares or envelopes in a lateral direction. Therefore, external circuit connections should be made to the porting block or square farthest from the actuator. This situation is shown in Fig. 17. If you were to "push" on the symbol representing the manual actuating lever, the two squares would slide to the right causing the arrow that indicates flow to align with the hydraulic lines leading to the rest of the circuit.

4. Arrows inside the squares or envelopes show direction of flow when that particular square is moved into working position. Look at Fig. 17 again. If the lever is pushed, the valve slides the left square or envelope into alignment with the hydraulic lines. The arrow indicates the direction of flow. In some circuits, fluid may be free to flow in either direction. In these cases a double-headed arrow is used.

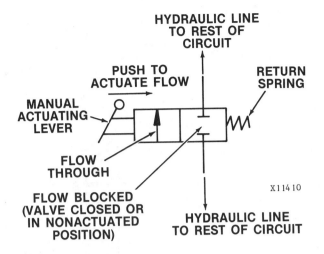

Fig. 17 — The Valve Must Slide To the Right in Order to Open the Hydraulic Circuit.

Two-Way Valves

In Fig. 18 we see a typical spool-type, two-way, normally closed valve both pictorially and by using ISO symbols. View 1 is a picture drawing of the valve in its normally closed or nonactuated state. The valve passage is blocked to oil flow. Graphically, the nonactuated valve is shown by a simple square (view 2) that indicates a blocked flow. The actuated valve is shown pictorially in view 3. Note that the ports are open to full flow. Graphically, the actuated valve is again shown by a square indicating full flow (view 4). When the two squares are put together (view 5), both flow functions of this 2-way, normally closed valve, are shown.

To complete the ISO symbol, the primary actuator (a manual lever in this case) is placed on one end and the return actuator (a spring) is placed on the other end (view 5).

Because the valve shown in Fig. 18 is normally closed the operator must hold it in the actuated or open state by keeping the handle firmly depressed. When the handle is released, spring tension returns the valve to its normally closed state.

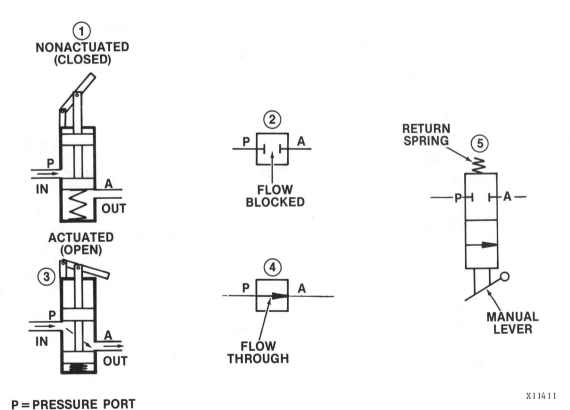

P = PRESSURE PORT
A = OUTLET PORT

Fig. 18 — The Operator Must Keep the Valve Lever Depressed to Keep the Hydraulic Oil Flowing Through the Valve.

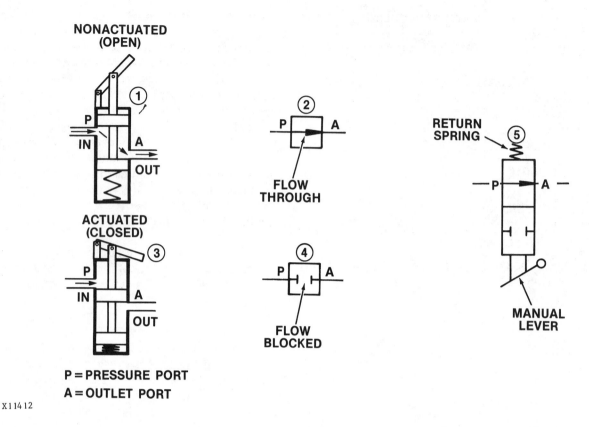

Fig. 19 — The Valve Lever Must Be Depressed In Order to Keep the Hydraulic Oil Blocked at the Valve.

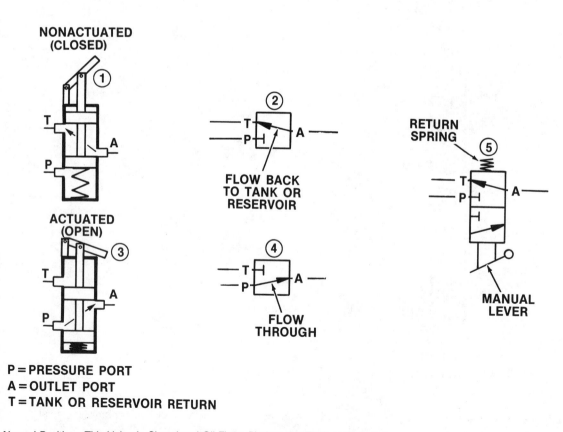

Fig. 20 — In Normal Positions This Valve is Closed and Oil Flows Back to the Tank or Reservoir.

In Fig. 19 we see a 2-way, normally open spool valve. It operates in reverse of the normally closed valve in that the spring returns it to the open position when it is nonactuated. Note that in the graphic illustration of this valve (view 5), the square indicating flow block is shown on the lever end of the symbol (remember rule 3).

Now, let's look at 3-way valves. They have three possible directions that the hydraulic fluid can flow.

Three-Way Valves

View 1 of Fig. 20 is a pictorial representation of a 3-way, normally closed valve in the nonactuated or closed state. Note that the pressure port is blocked. The flow of hydraulic fluid is only the return to the tank or reservoir. View 2 is the ISO symbol showing the same thing. Note the reversing arrow and the blocked pressure point.

View 3 of Fig. 20 shows the valve in the actuated or open position. Note that the pressure port is open to full flow to the outlet port. The return port to the tank or reservoir is blocked. View 4 is the ISO symbol. Note the arrow indicating flow from the pressure port to the outlet port. View 5 shows both squares or envelopes together with a manual lever and return spring added.

In Fig. 21 we see a 3-way, normally open spool valve. It operates in the reverse of the normally closed valve. They behave the same as the normally open and normally closed 2-way valves do.

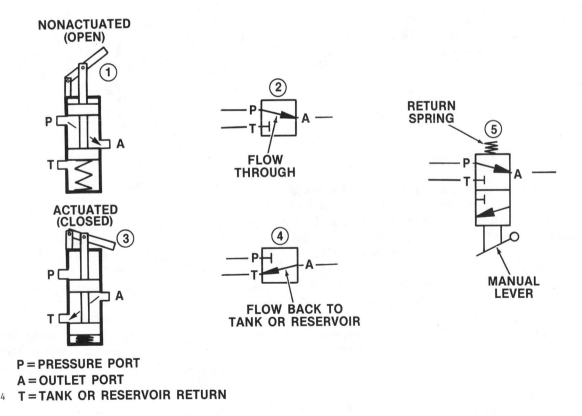

X11414

P = PRESSURE PORT
A = OUTLET PORT
T = TANK OR RESERVOIR RETURN

Fig. 21 — In Normal Position, This Valve is Open and Pressurized Oil Flows Through the Valve.

Three-Position Directional Control Valve

Although the valve shown in Fig. 22 looks very similar to the ones shown in Fig. 20 and 21, it is quite different. It has no normal position — it remains in whatever position the operator places it. It does not have an exhaust or reservoir return port, but it has two outlet ports.

A valve of this type is called a three-position, closed center, directional control valve. It permits the operation of two systems from one pump. When the handle is in the up position (view 1, Fig. 22), hydraulic pressure enters pressure port P and exits through outlet port A to one system (a cylinder, for example). Pressure is closed off to outlet port B. When the stem or handle is in the down position (view 3), pressure exits through port B to operate a second system (such as another cylinder). Port A is then closed to pressure. When the stem is centered (view 2), both outlet ports are closed with neither receiving pressurized hydraulic fluid.

That is why the valve is called a three-position, closed center, directional control valve. It can be placed in any one of three positions; pressure can be directed to either one of two systems; and when the stem is centered, the valve is closed. Notice that it takes three squares for the complete symbol of this valve — a square for each of the three positions.

Now let us look at a three-position, 4-way valve and see how we represent it with an ISO symbol.

Three-Position, 4-Way Valves

In Fig. 23 we see a three-position, 4-way valve with a closed center both pictorially and graphically. When the spool is shifted left, pressure enters through port P and is directed out port B to a cylinder. This allows port A to exhaust or drain back to the tank or reservoir through port T. When the spool is shifted right, the opposite happens — pressure from port P is routed to port A and port B exhausts to the tank (port T). When the spool is centered, all four ports are closed. Pressure is still present at port P but it can't go anywhere. The ISO symbols show this 4-way flow by using arrows to connect the ports.

The ISO symbol in Fig. 24 is the same valve shown in Fig. 23. The actuators and centering mechanisms have been added — in this case solenoids are the actuators and springs are the centering devices.

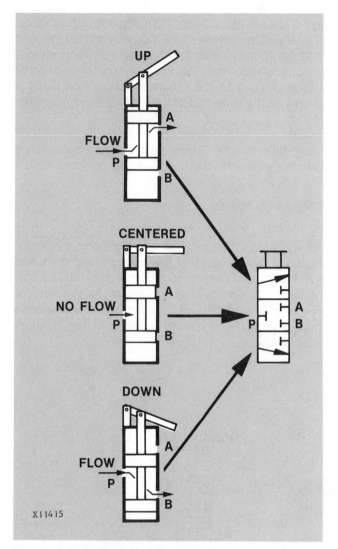

Fig. 22 — Two Systems Can Be Operated at the Same Time with This Three-Position, Closed Center Valve.

Flow Control Valves

A second type of control valves are needle valves and check valves.

Not all valves use the square as their basic symbol. Needle valves and check valves are types of valves that are represented by different symbols.

A flow control valve is shown in Fig. 25. It consists of a needle valve and check valve. Flow is controlled in one direction only. It is toward the right. Fluid flowing toward the left can pass in free flow through the internal check valve.

The ISO symbol of this valve (Fig. 25) has an arrow marking the direction of controlled flow and is an integral part of the symbol. An arrow (not a part of the symbol) shows free flow direction back through the check valve. Note also that the pressure port and free flow port are marked.

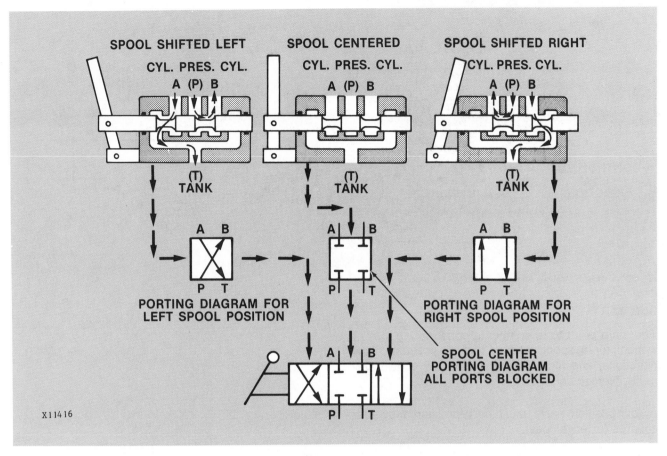

Fig. 23 — Pictorial and Symbolic Representation of a Three-Position, 4-Way Valve with a Normally Closed Center.

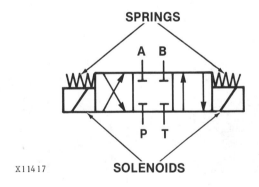

Fig. 24 — Electric Solenoids Can Be Used to Control Valves.

The valve in Fig. 25 is actually two commonly used hydraulic fluid flow restrictors. First, the adjustable needle valve is shown with the two arcs and the arrow through them. (This use of the arrow symbol shows an adjustment point whether it is used alone or with another symbol).

The second is the directional flow restrictor. The ball inside the arrowhead outline symbolizes that hydraulic fluid is free to flow from right to left in the valve (Fig. 25). Fluid cannot flow from left to right. This also is a common symbol that can be used alone. Remember, the two arrows shown outside of this valve are for clarification purposes only and are not part of the symbol.

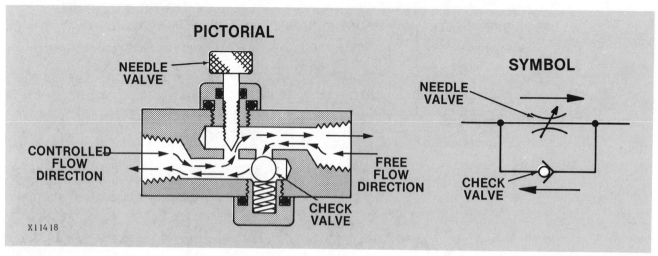

Fig. 25 — Needle Valves Also Are Fluid Control Valves

SUMMARY

The main point to remember in constructing an ISO symbol for a spool valve is that all positions, flow directions, and method of actuation are shown correctly. Remember:

A four-position valve must be represented by four squares (or envelopes).

A three-position valve must be represented by three squares.

A two-position valve must be represented by two squares.

One arrow must be used for each direction of flow. The ISO symbol for a 4-way valve would have four arrows, each indicating a direction of flow.

The method by which the valve is actuated must be shown. This includes a manual lever, solenoid, pilot, or other controls.

Flow control valves also restrict the direction of fluid flow, or the speed of fluid flow.

Pressure Control Valves

The third common type of valve is the pressure control valve. This type of valve can be a pressure **relief** or pressure **regulating** type.

The pressure relief valve helps control the hydraulic system pressure by opening if the system pressure gets too high. Hydraulic fluid drains back to the reservoir until the system pressure reaches the desired setting. Pressure relief valves are usually **adjustable** and can be pilot operated by back pressure in the system (Fig. 26).

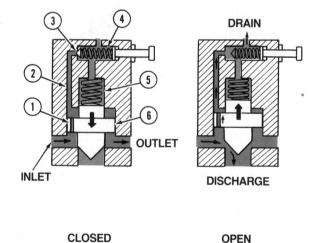

Fig. 26 — A Pressure Relief Valve Controls Hydraulic Pressure Internally in the System.

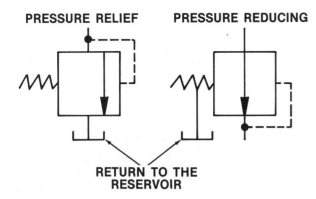

PRESSURE RELIEF   PRESSURE REDUCING

RETURN TO THE
RESERVOIR

Fig. 27 — A Pressure Control Valve Either Relieves A System
With Too Much Pressure or Helps Reduce the Pressure.

The symbol for a common type of pressure relief valve
is in Fig. 27.

The pressure reducing valve is usually a preset valve
that is intended to reduce hydraulic pressure in a
certain part of the hydraulic system. In Fig. 27 we can
see the symbolic representation. In Fig. 28 we can see
a pictorial view of a common pressure reducing valve.

Again we can see how the basic arrow and square
(envelope) can be used to represent a hydraulic valve.

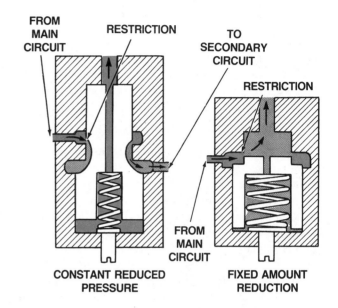

Fig. 28 — A Common Pressure Reducing Valve-Pictorial View.

PICTORIAL

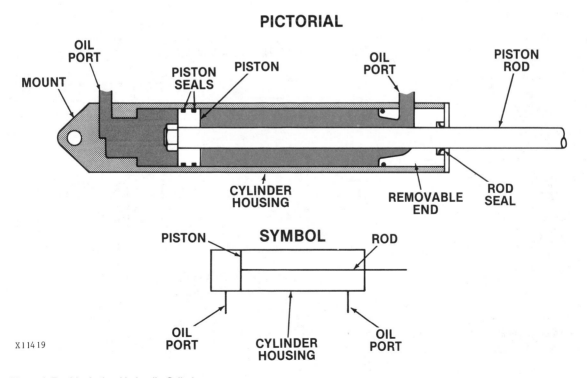

Fig. 29 — A Double-Acting Hydraulic Cylinder.

X11419

CYLINDERS

The ISO symbol for a hydraulic cylinder resembles a cylinder. In Fig. 29 we see both a pictorial representation and the ISO symbol for a typical double-acting hydraulic cylinder. (As you will recall from Chapter 4, a double-acting cylinder provides force in both directions of travel. Pressure is applied both the extend and then to retract it).

Clearly evident in the ISO symbol are the piston, rod, and cylinder housing. The ISO symbol does not indicate the piston seals, the type of clevis, or the internal configuration of the cylinder housing. This isn't necessary for a hydraulic circuit diagram.

In Fig. 30 we see a diagram that shows how two cylinders can be operated at the same time. In this figure the cylinders are labeled double-acting, but in an actual circuit diagram, nothing would be labeled. How would we know, then, that the cylinders are double-acting instead of single-acting? (Chapter 4 also describes single-acting.) Note that **both** ports on the cylinders can be connected to the pressure ports on the valves when the valves are actuated. (Remember rule 3 earlier in this chapter for the correct usage of ISO symbols.)

If the cylinders were single-acting, one of the ports on each cylinder would be a vent (not connected to the valves). Also, 4-way valves would not be necessary.

ACCUMULATORS

As we learned in Chapter 6, oil can be pumped into an accumulator and stored under pressure for use at a later time (Fig. 31).

Accumulators are not used in every hydraulic system. However, if a part such as a cylinder is in use less than 50 percent of the time a smaller pump and lower horsepower motor can be efficient with the aid of an accumulator. This reduces the pump and motor capacity and system cost.

In Fig. 32 we see a comparison of a conventional hydraulic circuit with an identical circuit using an accumulator. The accumulator is shown in the lower diagram along with its associated unloading valve and check valve. (Remember, a true hydraulic symbolic drawing does not label the parts of the system.)

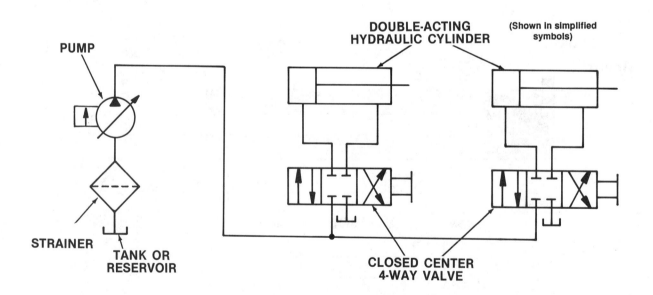

Fig. 30 — Two Double-Acting Cylinders Can Be Operated at the Same Time in the Same System.

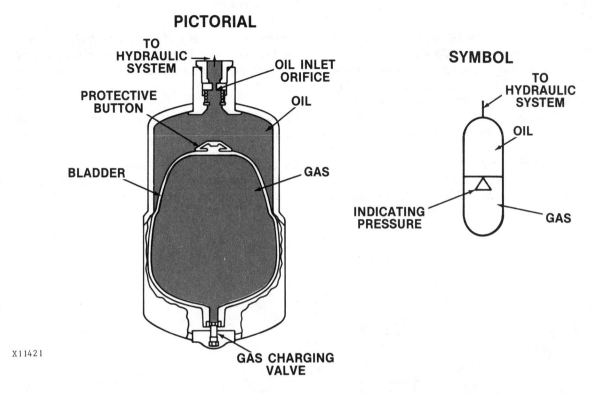

Fig. 31 — Accumulators Store Pressurized Hydraulic Fluid.

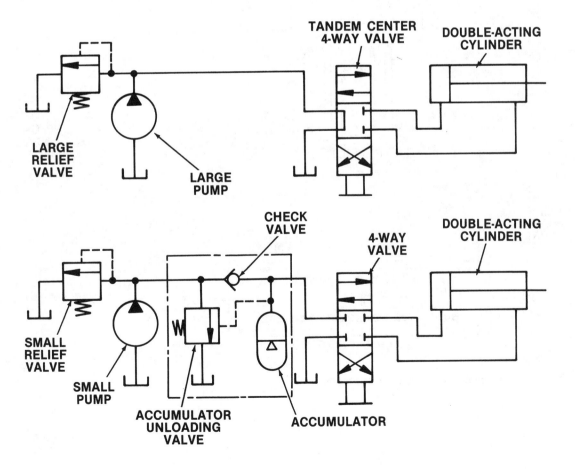

Fig. 32 — An Accumulator May be Economical if it Can Reduce the Size of the Pump and the Power Supply.

FILTERS

Filters, strainers, and other hydraulic fluid conditioners, all use the same symbol. In Fig. 33 we see a pictorial cross section view of a T-type and in-line filter along with the ISO symbol. The symbol is a diamond. The system hydraulic lines are connected to two of the corners. The filter or strainer element is represented by a dotted line connecting the other two corners.

Fig. 34 shows two circuits both using a suction strainer in the same location. The top diagram shows a micronic filter in the high-pressure pump line. The lower diagram shows the filter placed in the reservoir return line from the system relief valve.

OIL COOLERS

The two types of oil coolers (sometimes referred to as heat exchangers) that are widely used are air-to-oil and water-to-oil. Both are represented by the same ISO symbol.

In Fig. 35 we see a water-to-oil cooler along with the approved symbol.

Oil coolers will not stand high pressures. Therefore, they are installed in a low-pressure part of the system. This typical location is the reservoir return line. It is shown in Fig. 36.

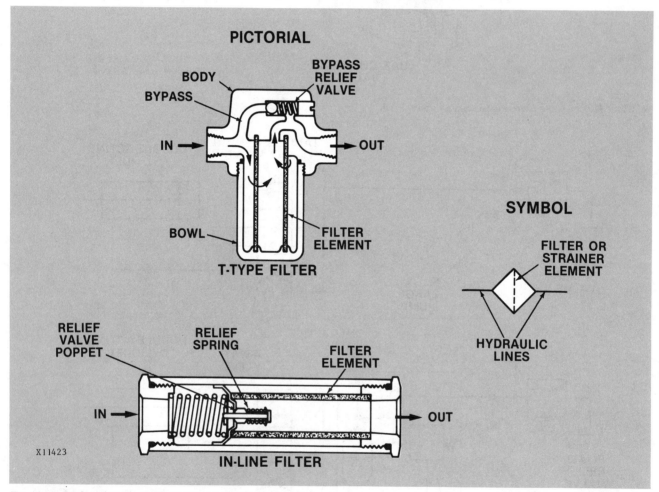

Fig. 33 — Oil Conditioners are Symbolized by Diamond Shapes.

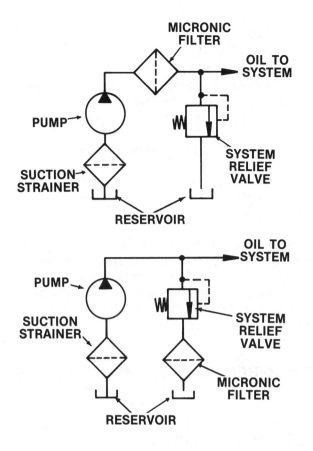

Fig. 34 — Filters Can Be Placed in Many Different Parts of the Hydraulic System.

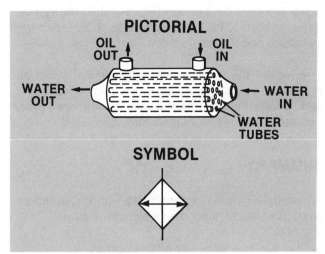

Fig. 35 — Water-to-Oil Coolers Transfer Heat from the Oil to the Water to Help Protect the Oil Quality.

HYDRAULIC RESERVOIRS

Hydraulic reservoirs are represented by the simplest ISO symbol (Fig. 37). You have been seeing this symbol on just about all the circuit diagrams in this chapter.

There is one main difference in its use and the other ISO symbols. Any given circuit diagram — Fig. 36, for example — may use the symbol several times, but in each instance, it represents the **same** reservoir. The symbol is used four times in Fig. 36 to represent different returns to the one reservoir. This prevents cluttering the diagram with so many return lines.

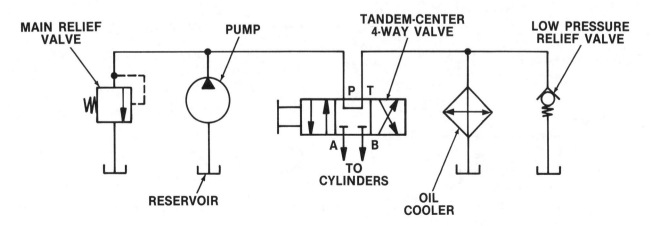

Fig. 36 — Oil Coolers are Installed in a Low Pressure Area of the Hydraulic System.

If you see two filter symbols (Fig. 34), then the system has two filters. The same applies to pumps, motors, cylinders, and valves.

The reason the symbol can be so simple is that all reservoirs are about the same. They may consist of a tank, gauge, baffle, air vent, intake screen, drain plug, and outlet and return lines as shown in Fig. 37.

SUMMARY

A hydraulic circuit diagram made with ISO symbols has certain advantages over a pictorial diagram.

• They are internationally understood

• They simplify design, fabrication, analysis, and service of hydraulic circuits

• They emphasize component function

• They show connections and flow paths

• They are easier to draw than the pictorial and cutaway drawings

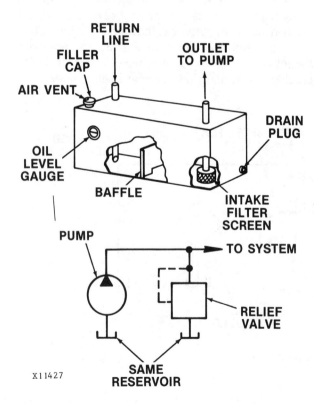

Fig. 37 — There Usually is Only One Hydraulic Fluid Reservoir in the Hydraulic System Even Though Many Symbols May Be Used.

To emphasis these points, let us look at Fig. 38 and Fig. 39. They are diagrams of the same circuit. One is pictorial, and the other is an ISO symbol diagram. Note how much simpler and easier to draw Fig. 39 is over Fig. 38. This is why the large majority of system diagrams that you will see will be drawn with symbols.

You must be knowledgeable about ISO symbols.

To help your learn ISO symbols, study Fig. 40. It contains all the basic symbols that you need to help interpret hydraulic circuit diagrams.

All symbolic diagrams are drawn with the basic forms — the circle, square, diamond, and rectangle. They are further defined with basic marks including arrows, lines, arrowheads, and arcs. With these basic concepts you will be able to decode even the most complex symbolic diagrams.

HOW TO READ THE OIL FLOW DIAGRAMS

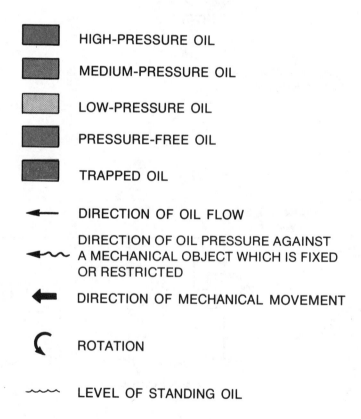

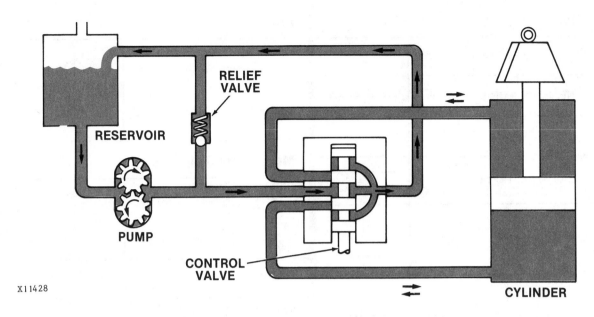

X11428

Fig. 38 — A Pictorial Diagram of a Hydraulic System.

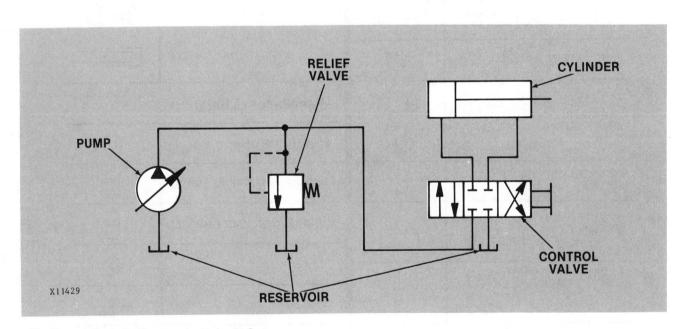

X11429

Fig. 39 — A Symbolic Diagram of a Hydraulic System.

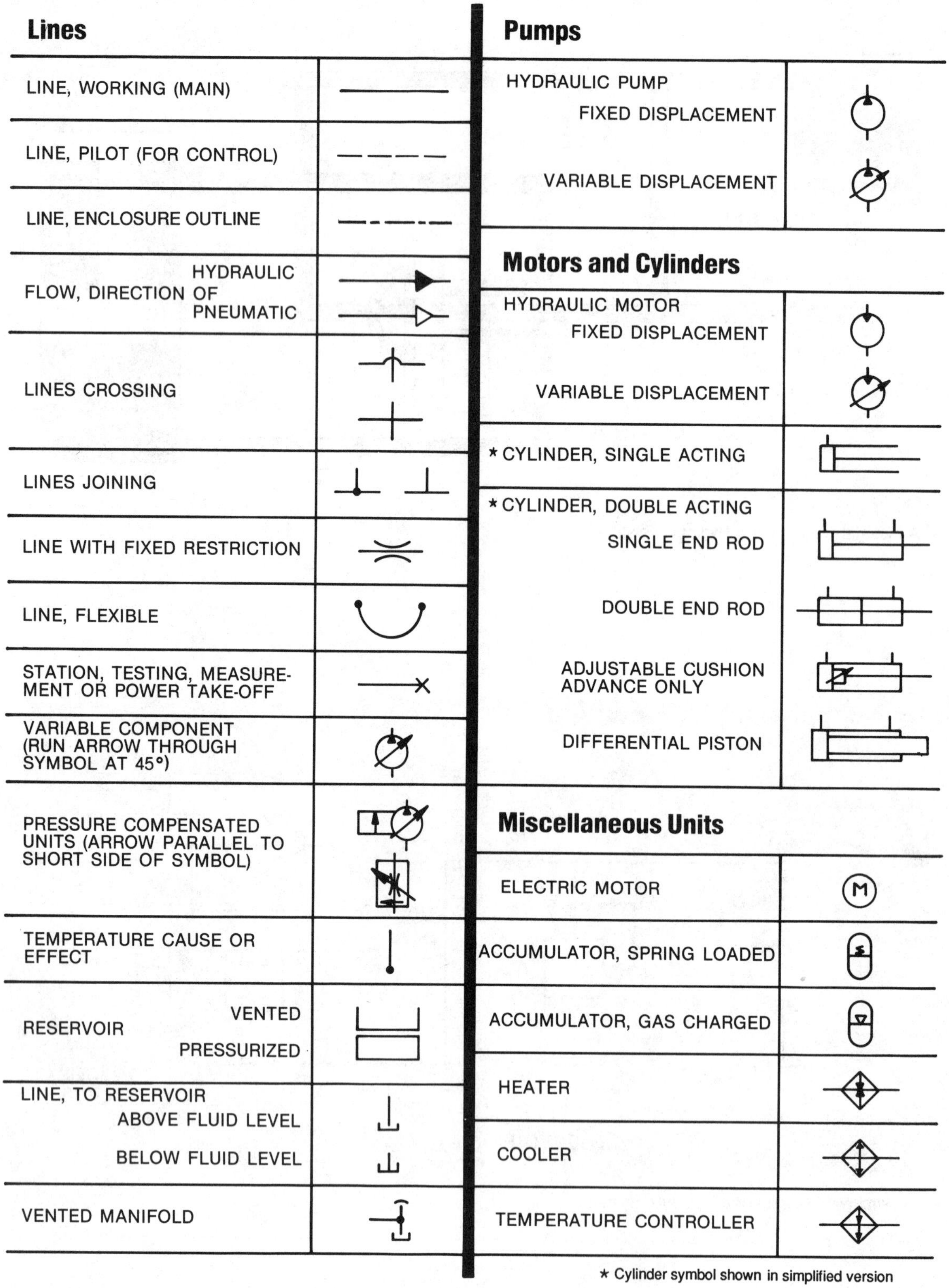

Lines

LINE, WORKING (MAIN)	
LINE, PILOT (FOR CONTROL)	
LINE, ENCLOSURE OUTLINE	
FLOW, DIRECTION OF — HYDRAULIC / PNEUMATIC	
LINES CROSSING	
LINES JOINING	
LINE WITH FIXED RESTRICTION	
LINE, FLEXIBLE	
STATION, TESTING, MEASUREMENT OR POWER TAKE-OFF	
VARIABLE COMPONENT (RUN ARROW THROUGH SYMBOL AT 45°)	
PRESSURE COMPENSATED UNITS (ARROW PARALLEL TO SHORT SIDE OF SYMBOL)	
TEMPERATURE CAUSE OR EFFECT	
RESERVOIR — VENTED / PRESSURIZED	
LINE, TO RESERVOIR — ABOVE FLUID LEVEL / BELOW FLUID LEVEL	
VENTED MANIFOLD	

Pumps

HYDRAULIC PUMP FIXED DISPLACEMENT	
VARIABLE DISPLACEMENT	

Motors and Cylinders

HYDRAULIC MOTOR FIXED DISPLACEMENT	
VARIABLE DISPLACEMENT	
* CYLINDER, SINGLE ACTING	
* CYLINDER, DOUBLE ACTING SINGLE END ROD	
DOUBLE END ROD	
ADJUSTABLE CUSHION ADVANCE ONLY	
DIFFERENTIAL PISTON	

Miscellaneous Units

ELECTRIC MOTOR	
ACCUMULATOR, SPRING LOADED	
ACCUMULATOR, GAS CHARGED	
HEATER	
COOLER	
TEMPERATURE CONTROLLER	

* Cylinder symbol shown in simplified version

Fig. 40A — ISO Hydraulic Symbols

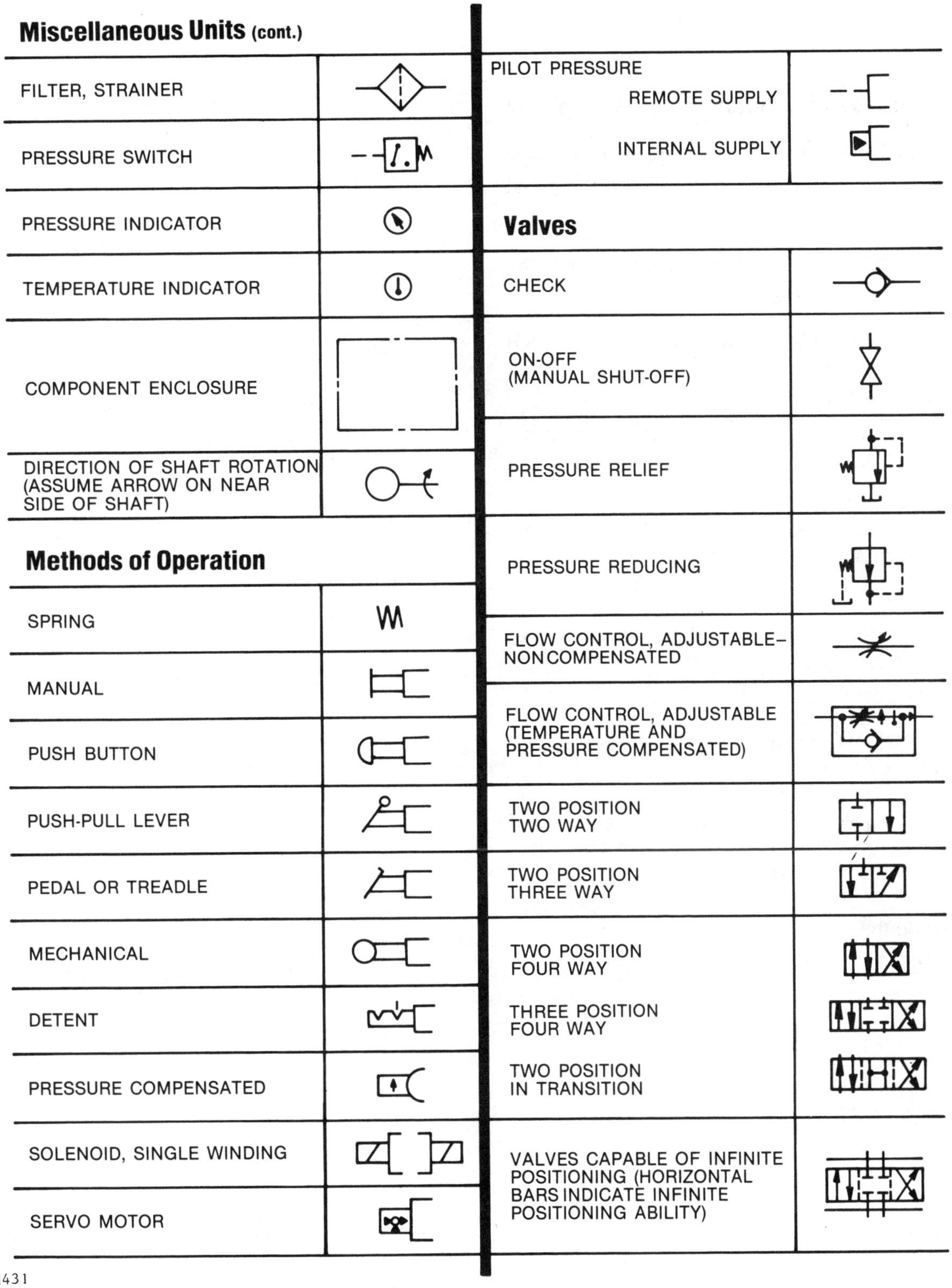

Miscellaneous Units (cont.)

FILTER, STRAINER	
PRESSURE SWITCH	
PRESSURE INDICATOR	
TEMPERATURE INDICATOR	
COMPONENT ENCLOSURE	
DIRECTION OF SHAFT ROTATION (ASSUME ARROW ON NEAR SIDE OF SHAFT)	

Methods of Operation

SPRING	
MANUAL	
PUSH BUTTON	
PUSH-PULL LEVER	
PEDAL OR TREADLE	
MECHANICAL	
DETENT	
PRESSURE COMPENSATED	
SOLENOID, SINGLE WINDING	
SERVO MOTOR	

PILOT PRESSURE

REMOTE SUPPLY

INTERNAL SUPPLY

Valves

CHECK	
ON-OFF (MANUAL SHUT-OFF)	
PRESSURE RELIEF	
PRESSURE REDUCING	
FLOW CONTROL, ADJUSTABLE— NON COMPENSATED	
FLOW CONTROL, ADJUSTABLE (TEMPERATURE AND PRESSURE COMPENSATED)	
TWO POSITION TWO WAY	
TWO POSITION THREE WAY	
TWO POSITION FOUR WAY	
THREE POSITION FOUR WAY	
TWO POSITION IN TRANSITION	
VALVES CAPABLE OF INFINITE POSITIONING (HORIZONTAL BARS INDICATE INFINITE POSITIONING ABILITY)	

X11431

Fig. 40B — ISO Hydraulic Symbols

TEST YOURSELF

QUESTIONS

1. What are three advantages in using ISO symbols?

2. The circle is used to represent both pumps and motors. How do you tell which is which on a circuit diagram?

3. What does the large arrow running through the circle symbol at a 45 degree angle tell you?

4. (True or false) A 4-way valve is always represented by four squares.

5. A three-position valve is represented by _____ squares.

6. In an ISO symbol diagram, connections to the rest of the circuit from a three-position valve (three squares) are made from how many of the squares? Why?

7. Connections from a valve to the rest of the circuit are made from the square that indicates the _____ state of the valve. (actuated, nonactuated)

8. (True or false) Filters and strainers are represented by the same symbol.

9. What hydraulic component may be represented several times on a diagram even though the actual system will have only one?

10. What do the initials ISO stand for?

SAFETY RULES FOR HYDRAULICS / CHAPTER 14

INTRODUCTION

In the very beginning of this book, we explained the high forces and pressures that can be generated in the hydraulic system. These can cause serious injuries and even death, if reasonable care is not exercised. The following information is designed to alert you to the potential hazards, and to keep you safe if you follow the instructions.

HYDRAULIC SYSTEMS

Hydraulic systems store energy. Hydraulic systems must confine fluid under high pressure often higher than 2,000 pounds per square inch.

A lot of energy may be stored in a hydraulic system, and because there is often no visible motion, operators do not recognize it as a potential hazard. Carelessly servicing, adjusting, or replacing parts can result in serious injury. Fluid under pressure attempts to escape (Fig. 2). In doing so it can do helpful work, or it can be harmful.

 CAUTION: Never service or adjust systems under pressure.

Adjusting and removing components when hydraulic fluid is under pressure can be hazardous (Fig. 3). Imagine attempting to remove a faucet from your kitchen sink without relieving the water pressure. You'd get a face full of water! It is much more dangerous with hydraulic systems. Instead of just getting wet from water at 40 psi, you would be seriously injured by oil under 2,000psi, or more. You could be injured by the hot, high pressure spray of fluid and by the part you are removing when it is thrown at you (Fig. 3).

Fig. 1 — Always Exercise Safety

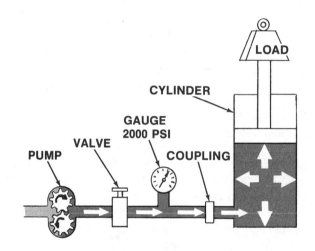

Fig. 2 — Hydraulic Fluid Under Pressure Attempts To Escape Or Move To A Point Of Lower Pressure

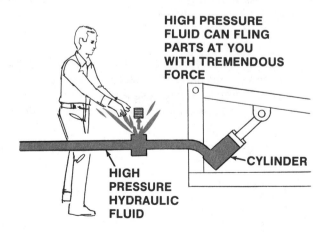

Fig. 3 — Always Relieve Hydraulic Pressure Before Adjusting Hydraulic Fittings. You Could Be Injured By A Hot, High Pressure Spray of Hydraulic Fluid Or By A Part Flung At You

AVOID HIGH-PRESSURE FLUIDS

Escaping fluid under pressure can penetrate the skin causing serious injury.

Avoid the hazard by relieving pressure before disconnecting hydraulic or other lines. Tighten all connections before applying pressure.

Search for leaks with a piece of cardboard. Protect hands and body from high pressure fluids.

If an accident occurs, see a doctor immediately. Any fluid injected into the skin must be surgically removed within a few hours or gangrene may result. Doctors unfamiliar with this type of injury should consult with medical sources to obtain pertinent information.

Fig. 4 — Avoid High-Pressure Fluids

Pinhole Leaks Can Be Dangerous

If liquid, under pressure, escapes through an extremely small opening, it comes out as a fine stream (Fig. 5). The stream is called a pinhole leak. Pinhole leaks in hydraulic systems are hard to see and they can be very dangerous. High-pressure streams of oil from a pinhole leak can penetrate human flesh. Hydraulic systems often have pressures over 2,000psi. That's higher than the pressure in hydraulic syringes used to give injections. Injuries caused by fluids (oil) injected into human flesh can be very serious. Consult a physician immediately, if you believe that you have been injured by a pinhole stream.

 CAUTION: Never try to detect the pinhole leak by running your hand over the area where you suspect the leak. Always use a piece of cardboard (Fig. 5). Also, wear safety glasses or a face shield.

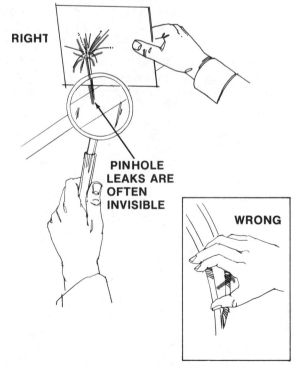

RIGHT

PINHOLE LEAKS ARE OFTEN INVISIBLE

WRONG

Fig. 5 — The Jet Stream Or Mist From A Pinhole Leak In A Hydraulic System Can Penetrate Your Skin — Don't Touch It!

AVOID HAZARD OF STORED ENERGY FROM HYDRAULIC ACCUMULATOR.

Some hydraulic systems have accumulators to store energy. They may also be used to absorb shock loads and to maintain a constant pressure in the system. Recognize that the accumulators and the entire hydraulic system may have energy stored in it, if the pressure has not been relieved. Even though the pump may be stopped, or an implement has been disconnected from the tractor, energy is stored in the accumulator unless the pressure was relieved before shutdown. The nitrogen is under pressure, so the hydraulic fluid is also under pressure (Fig. 6).

⚠ **CAUTION: Observe these basic safety considerations for hydraulic accumulators:**

1. *Recognize accumulators as sources of stored energy.*

2. *Relieve all hydraulic system pressure before adjusting or servicing any part of an accumulator system.*

3. *Relieve all hydraulic pressure before leaving a machine unattended* for the safety of others as well as for you.

4. *Make sure pneumatic accumulators are properly charged with the proper inert gas (usually nitrogen).* A pneumatic accumulator without gas is a potential "bomb" when charged only with oil.

5. *Read and follow manufacturer's instructions for servicing accumulators.*

Generally, manufacturers recommend only authorized dealers service gas charged accumulators. Read and follow the manufacturer's instructions thoroughly.

AVOID HAZARD OF TRAPPED OIL AND THERMAL EXPANSION

Another hazard with trapped oil (Fig. 7) is heat. Heat from the sun can expand the hydraulic oil and increase pressure. The pressure can blow seals and move parts of an implement or machine.

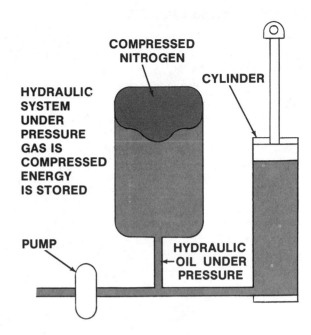

Fig. 6 — Hydraulic Accumulators Store Energy

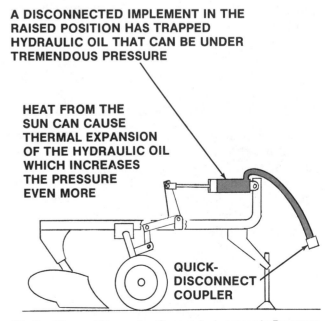

Fig. 7 — Thermal Expansion Causes An Increase In Pressure

CONNECT LINES CORRECTLY

The movement of hydraulic components should correspond to the movement of the controls. If a control lever is placed in the "raise" position, the function should raise; or if the steering wheel is turned to the left, the wheels should turn left.

Wrong hook-up of lines or hoses (Fig. 8) will cause the reverse of the intended action. This may result in an "element of surprise" and could lead to serious injuries. Carefully test the machine after each repair.

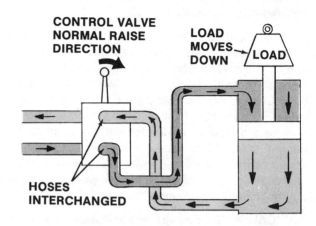

Fig. 8 — Avoid Incorrect Hose Connection

AVOID CRUSH POINTS

There are crush points between two objects that move toward each other or one object moving toward a stationary object.

Many of the machine movements are trigered by hydraulic action. There are, for instance, several dangerous crush points on hydraulic hitches.

Another dangerous crush point is between tires or frame parts on tractors with articulated steering (Fig. 9). There can be immediate movement upon start-up of the tractor when the steering wheel isn't even moved. Hydrostatic steering systems are very sensitive. NEVER stand between the tires during start-up or any time the engine is running. NEVER allow anyone else to stand there.

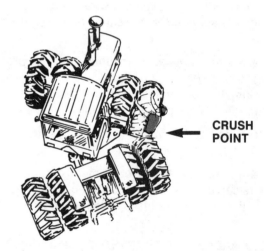

Fig. 9 — Four-Wheel Drive Tractors With Articulated Steering Can Create A Crush Point

AVOID HEATING NEAR PRESSURIZED FLUID LINES

Flammable spray can be generated by heating near pressurized fluid lines (Fig. 10), resulting in severe burns to yourself and bystanders. Do not heat by welding, soldering, or using a torch near pressurized fluid lines or other flammable materials. Pressurized lines can be accidentally cut when heat goes beyond the immediate flame area.

Fig. 10 — Avoid Heating Near Pressurized Fluid Lines

GENERAL SAFETY INSTRUCTIONS

On the previous pages, you have been given safety information that is specific to the hydraulic system. But when working on the hydraulic system, it is often necessary to remove and replace non-hydraulic components. Therefore, total care has to be exercised in all service activities.

The following safety instructions, if followed, are designed to keep you safe. Make it a practice to work safely. It's the right thing to do.

Fig. 11 — Always Exercise Safety

RECOGNIZE SAFETY INFORMATION

This is the safety-alert symbol. When you see this symbol on your machine, in this manual, or the machine's manual, be alert to the potential for personal injury.

Follow recommended precautions and safe operating practices.

Fig. 12 — Safety Alert Symbol

UNDERSTAND SIGNAL WORDS

A signal word—DANGER, WARNING, or CAUTION—is used with the safety-alert symbol. DANGER identifies the most serious hazards.

DANGER or WARNING safety signs are located near specific hazards. General precautions are listed on CAUTION safety signs. CAUTION also calls attention to safety messages in this manual, and the machine's manual.

Fig. 13 — Signal Words

FOLLOW SAFETY INSTRUCTIONS

Carefully read all safety messages in this manual and on the machine's safety signs. No two machines are exactly alike, so don't take any chances. Become familiar with the safety instructions and follow them. Also insist that those working with you follow the instructions. It will keep you and them safe.

Fig. 14 — Follow Safety Instructions

PRACTICE SAFE MAINTENANCE

Understand the service procedure before doing the work. Keep the area clean and dry.

Never lubricate or service the machine while it is moving. Keep hands, feet, and clothing from power driven parts. Disengage all power and operate the controls to relieve pressure. Lower the equipment to the ground. Stop the engine. Remove the key. Allow the machine to cool.

Securely support any machine elements that must be raised for service work.

Keep all parts in good condition and properly installed. Fix damage immediately. Replace worn or broken parts. Remove any buildup of grease, oil, or debris.

Disconnect the battery ground cable (-) before making adjustments on the electrical systems or welding on the machine.

Fig. 15 — Practice Safe Maintenance

WORK IN CLEAN AREA

Before starting a job:
• Clean the work area and the machine.
• Make sure you have all the necessary tools to do your job.
• Have the right parts on hand.
• Read all the instructions thoroughly; do not attempt shortcuts.

Fig. 16 — Work In Clean Area

PREPARE FOR EMERGENCIES

Be prepared if a fire starts.

Keep a first aid kit and fire extinguisher handy.

Keep emergency numbers for doctors, ambulance service, hospital, and fire department near your telephone.

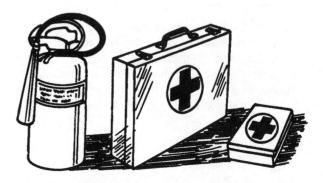

Fig. 17 — Prepare For Emergencies

PARK MACHINE SAFELY

Before working on the machine:
• Lower all equipment to the ground.
• Stop the engine and remove the key.
• Disconnect the battery ground strap.
• Hang a "DO NOT OPERATE" tag in operator station.

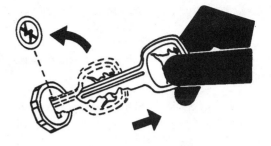

Fig. 18 — Park Machine Safely

SERVICE MACHINES SAFELY

Tie long hair behind your head. Do not wear a necktie, scarf, loose clothing, or necklace when you work near machine tools or moving parts. If these items were to get caught, severe injury could result.

Remove rings and other jewelry to prevent electrical shorts and entanglement in moving parts.

Fig. 19 — Service Machine Safely

USE PROPER TOOLS

Use tools appropriate to the work. Makeshift tools and procedures can create safety hazards.

Use power tools only to loosen threaded parts and fasteners.

For loosening and tightening hardware, use the correct size tools. DO NOT use U.S. measurement tools on metric fasteners. Avoid bodily injury caused by slipping wrenches.

Fig. 20 — Use Proper Tools

HANDLE FLUIDS SAFELY—AVOID FIRES

When you work around fuel, do not smoke or work near heaters or other fire hazards.

Store flammable fluids away from fire hazards. Do not incinerate or puncture pressurized containers.

Make sure machine is clean of trash, grease, and debris.

Do not store oily rags; they can ignite and burn spontaneously.

Fig. 21 — Avoid Fires

PREVENT MACHINE RUNAWAY

Avoid possible injury or death from machinery runaway.

Do not start the engine by shorting across the starter terminals. The machine will start in gear if normal circuitry is bypassed.

NEVER start the engine while standing on the ground. Start the engine only from the operator's seat, with the transmission in neutral or park.

Fig. 22 — Prevent Machine Runaway

DIPSOSE OF FLUIDS PROPERLY

Improperly disposing of fluids can harm the environment and ecology. Before draining any fluids, find out the proper way to dispose of waste from your local environmental agency.

Use proper containers when draining fluids. Do not use food or beverage containers that may mislead someone into drinking from them.

DO NOT pour oil into the ground, down a drain, or into a stream, pond, or lake. Observe relevant environmental protection regulations when disposing of oil, fuel, coolant, brake fluid, filters, batteries, and other harmful waste.

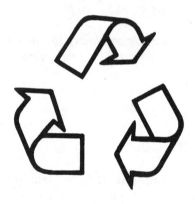

Fig. 23 — Dispose of Fluids Properly

TEST YOURSELF

QUESTIONS

1. What does the safety alert symbol indicate?

2. What are the three safety signal words?

3. What must be used to locate a high pressure leak?

4. Hydraulic accumulators store.

5. What is a crush point?

6. What four things must be done before any service is performed?

DEFINITIONS OF TERMS AND SYMBOLS

A

ACCUMULATOR—A container which stores fluids under pressure as a source of hydraulic power. It may also be used as a shock absorber.

ACTUATOR—A device which converts hydraulic power into mechanical force and motion. (Examples: hydraulic cylinders and motors.)

B

BLEED—The process by which air is removed from a hydraulic system.

BYPASS—A secondary passage for fluid flow.

C

CAM LOBE MOTOR—A hydraulic radial piston motor in which rotational force is created by the outward movement of the pistons against the lobes of a stationary cam.

CAVITATION—A phenomenon which occurs when the pressure at a point in a hydraulic system is lowered below the vapor pressure of the oil in the system. This allows bubbles of oil vapor to form in the oil. If this occurs at the pump inlet, the quick pressure rise inside the pump forces these bubbles to collapse violently. This can cause erosion of metal parts, noise and vibration.

CIRCUIT—A series of component parts connected to each other by fluid lines or passages. Usually part of a "system".

CLOSED CENTER SYSTEM—A hydraulic system in which the control valves are closed during neutral, stopping oil flow. Flow in this system is varied, but pressure remains constant.

CONTROLLER—A microprocessor that controls electro-hydraulic valve functions.

COOLER (Oil)—A heat exchanger which removes heat from a fluid. (See "Heat Exchanger.")

COUPLER—A device to connect two hoses or lines, or to connect hoses to valve receptacles.

CUSHION—A device sometimes built into the end of a cylinder which restricts outlet flow and thereby slows down the piston.

CYCLE—A single complete operation of a component which begins and ends in a neutral position.

CYLINDER—A device for converting fluid power into linear or circular motion. An "actuator". Basic design types include piston and vane units.

Double-Acting Cylinder—A cylinder in which fluid force can be applied to the movable element in either direction.

Piston-Type Cylinders—A cylinder which uses a sliding piston in a housing to produce straight movement.

Rotary Cylinders—A cylinder in which fluid force is applied to produce circular motion.

Single-Acting Cylinder—A cylinder in which fluid force can be applied to the movable element in only one direction.

Vane-Type Cylinder—A cylinder which uses a turning vane in a circular housing to produce rotary movement.

D

DISPLACEMENT—The volume of oil displaced by one complete stroke or revolution (of a pump, motor, or cylinder).

DRIFT—Motion of a cylinder or motor due to internal leakage past components in the hydraulic system.

E

ENERGY—Three types of energy are available in modern hydraulics (of the normal hydrostatic type):

1. *Potential Energy*—Pressure energy. The *static* energy of oil which is standing but is pressurized and ready to do work. Example: oil in a loaded accumulator.

2. *Heat Energy*—Friction or resistance to flow. (An energy *loss* in terms of output.) Example: friction between moving oil and the confines of lines or passages produces heat energy.

3. *Kinetic Energy*—The energy of the moving liquid. Varies with the velocity (speed) of the liquid.

F

FILTER (OIL)—A device which removes solids from a fluid.

FLOW METER—A testing device which gauges either flow rate, total flow, or both.

FLOW RATE—The volume of fluid passing a point in a given time.

FLUID POWER—Energy transmitted and controlled through use of a pressurized fluid.

FORCE—A push or pull acting upon a body. In a hydraulic cylinder, it is the product of the pressure on the fluid, multiplied by the effective area of the cylinder piston. It is measured in pounds or tons.

FRICTION—The resistance to fluid flow in a hydraulic system. (An energy loss in terms of power output.)

H

HEAT EXCHANGER—A device which transfers heat through a conducting wall from one fluid to another. (See "Cooler, (Oil)".)

HORSEPOWER — The work produced per unit of time.

HOSE—A flexible line.

HYDRAULICS—The engineering science of liquid pressure and flow. (In this manual, our main interest is in *oil* hydraulics as applied to produce work in linear and rotary planes.)

Hydrodynamics—The engineering science of the *energy* of liquid pressure and flow.

Hydrostatics—The engineering science of the energy of liquids *at rest*. (All the systems covered in this manual operate on the *hydrostatic* principle.)

I

INERT GAS—A non-explosive gas.

L

LINE—A tube, pipe, or hose for conducting a fluid.

M

MANIFOLD—A fluid conductor which provides many ports.

MOTOR (Hydraulic)—A device for converting fluid energy into mechanical force and motion—usually rotary motion. Basic design types include gear, vane, and piston units.

O

OPEN CENTER SYSTEM—A hydraulic system in which the control valves are open to continuous oil flow, even in neutral. Pressure in this system is varied, but flow remains constant.

ORIFICE—A restricted passage in a hydraulic circuit. Usually a small drilled hole to limit flow or to create a pressure differential in a circuit.

O-RING — A static and/or dynamic seal for curved or circular mating surfaces.

P

PACKING—Any material or device which seals by compression. Common types are U-packings, V-packings, "Cup" packings, and O-rings.

PIPE—A line whose outside diameter is standardized for threading.

PISTON—A cylindrical part which moves or reciprocates in a cylinder and transmits or receives motion to do work.

PORT—The open end of a fluid passage. May be within or at the surface of a component.

POUR POINT—The lowest temperature at which a fluid will flow under specific conditions.

POWER BEYOND—An adapting sleeve which opens a passage from one circuit to another. Often installed in a valve port which is normally plugged.

PRESSURE—Force of a fluid per unit area, usually expressed in pounds per square inch (psi).

Back Pressure—The pressure encountered on the return side of a system.

Breakout Pressure—The minimum pressure which starts moving an actuator.

Cracking Pressure—The pressure at which a relief valve, etc., begins to open and pass fluid.

Differential Pressure—The difference in pressure between any two points in a system or a component. (Also called a "pressure drop.")

Full-Flow Pressure—The pressure at which a valve is wide open and passes its full flow.

Operating Pressure—The pressure at which a system is normally operated.

Pilot Pressure—Auxiliary pressure used to actuate or control a component.

Rated Pressure—The operating pressure which is recommended for a component or a system by the manufacturer.

Static Pressure—The pressure in a fluid at rest. (A form of "potential energy.")

Suction Pressure—The absolute pressure of the fluid at the inlet side of the pump.

Surge Pressure—The pressure changes caused in a circuit from a rapidly accelerated column of oil. The "surge" includes the span of these changes, from high to low.

System Pressure—The pressure which overcomes the total resistances in a system. It includes all losses as well as useful work.

Working Pressure—The pressure which overcomes the resistance of the working device.

PULSATION—Repeated small fluctuation of pressure within a circuit.

PUMP—A device which converts mechanical force into hydraulic fluid power. Basic design types are gear, vane, and piston units.

Fixed Displacement Pump—A pump in which the output per cycle *cannot* be varied.

Variable Displacement Pump—A pump in which the output per cycle *can* be varied.

R

REGENERATIVE CIRCUIT — A circuit in which pressure fluid discharged from a component is returned to the system to reduce flow input requirements. Often used to speed up the action of a cylinder by directing discharged oil from the rod end to the piston end.

REMOTE—A hydraulic function such as a cylinder which is separate from its supply source. Usually connected to the source by flexible hoses.

RESERVOIR—A container for keeping a supply of working fluid in a hydraulic system.

RESTRICTION—A reduced cross-sectional area in a line or passage which normally causes a pressure drop. (Examples: pinched lines or clogged passages, or an orifice designed into a system.)

S

SOLENOID—An electro-magnetic device which positions a hydraulic valve.

STARVATION—A lack of oil in vital areas of a system. Often caused by plugged filters, etc.

STRAINER—A coarse filter.

STROKE—
1. The length of travel of a piston in a cylinder.
2. Sometimes used to denote the changing of the displacement of a variable delivery pump.

SURGE—A momentary rise of pressure in a hydraulic circuit.

SYMBOLS, SCHEMATIC — Used as a short-hand representation on drawings to represent hydraulic system components.

SYSTEM—One or more series of component parts connected to each other. Often made up of two or more "circuits".

T

THERMAL EXPANSION—Expansion of the fluid volume due to heat.

TORQUE—The turning effort of a hydraulic motor or rotary cylinder. Usually given in inch-pounds (in-lbs) or foot-pounds (ft-lbs).

TUBE—A line whose size is its outside diameter.

V

VALVE—A device which controls either 1) pressure of fluid, 2) direction of fluid flow, or 3) rate of flow.

Bypass Flow Regulator Valve—A valve which regulates the flow to a circuit at a constant volume, dumping excess oil.

Check Valve—A valve which permits flow in only one direction.

Closed Center Valve—A valve in which inlet and outlet ports are closed in the neutral position, stopping flow from pump.

Directional Control Valve—A valve which directs oil through selected passages. (Usually a spool or rotary valve design.)

Electro-Hydraulic Valve—A valve that is opened and closed by a solenoid.

Flow Control Valve—A valve which controls the rate of flow. (Sometimes called a "volume control valve.")

Flow Divider Valve—A valve which divides the flow from one source into two or more branches. (Includes "priority" and "proportional" types.)

Needle Valve—A valve with an adjustable tapered point which regulates the rate of flow.

Open Center Valve—A valve in which the inlet and outlet ports are open in the neutral position, allowing a continuous flow of oil from pump.

Pilot Valve—A valve used to operate another valve or control.

4

Pilot Operated Valve—A valve which is actuated by a pilot valve.

Poppet Valve—A valve design in which the seating element pops open to obtain free flow in one direction and immediately reseats when flow reverses.

Pressure Control Valve—A valve whose primary function is to control pressure. (Includes relief valves, pressure reducing or sequencing valves, and unloading valves.)

Pressure Reducing Valve—A pressure control valve which limits outlet pressure.

Pressure Sequence Valve—A pressure control valve which directs flow in a preset sequence.

Priority Flow Divider Valve—A valve which directs oil to one circuit at a fixed rate and dumps excess flow into another circuit.

Proportional Flow Divider Valve—A valve which directs oil to all its circuits at all times.

Relief Valve—A valve which limits the pressure in a system, usually by releasing excess oil.

Rotary Directional Valve—A valve designed in a cylindrical shape. When the valve is turned, it opens and closes drilled passages to direct oil.

Selector Valve—A valve which selects one of two or more circuits in which to direct oil, usually operated manually.

Shuttle Valve—A connecting valve which selects one of two or more circuits because of flow or pressure changes in these circuits.

Shutoff Valve—A valve which operates fully open or fully closed.

Spool Directional Valve—A valve designed as a spool which slides in a bore, opening and closing passages.

Thermal Relief Valve—A valve which limits the pressure in a system caused by heat expansion of oil.

Two-, Three-, Four-, or Six-Way Valve—A valve having 2, 3, 4, or 6 ports for direction of oil flow.

Unloading Valve—A valve which allows a pump to operate at minimum load by dumping the pump's excess oil at a low pressure.

Volume Control Valve—A valve which controls the rate of flow. Includes flow control valves, flow divider valves, and bypass flow regulators.

V

VALVE STACK—A series of control valves in a stack with common end plates and a common oil inlet and outlet.

VELOCITY—The *distance* which a fluid travels per unit time. Usually given as feet per second.

VENT—An air breathing device in a fluid reservoir.

VISCOSITY—The measure of resistance of a fluid to flow.

VOLUME—The amount of fluid flow per unit time. Usually given as gallons per minute (gpm).

ABBREVIATIONS

ANSI—American National Standards Institute

ASAE—American Society of Agricultural Engineers (sets standards for many hydraulic components for agricultural use)

°F—degrees Fahrenheit (of temperature)

ft-lbs—foot-pounds (of torque or turning effort)

gpm—gallons per minute (of fluid flow)

hp—horsepower

I.D.—inside diameter (as of a hose or tube)

ISO—International Standards Organization

O.D.—outside diameter (as of a hose or tube)

psi—pounds per square inch (of pressure)

rpm—revolutions per minute

SAE—Society of Automotive Engineers (sets standards for many hydraulic components)

METRIC CHART

Metric to English

LENGTH
1 millimeter = 0.039 37 inches ....in
1 meter = 3,281 feet ....ft
1 kilometer = 0.621 miles ....mi

AREA
1 meter² = 10.76 feet² ....ft²
1 hectare = 2.471 acres ....acre
  (hectare = 10,000 m²)

MASS (WEIGHT)
1 kilogram = 2.205 pounds ....lb
1 tonne (1000 kg) = 1.102 short ton .....sh tn

VOLUME
1 meter³ = 35,31 foot³ ....ft³
1 meter³ = 1.308 yard³ ....yd³
1 meter³ = 28.38 bushel ....bu
1 liter = 0.028 38 bushel ....bu
1 liter = 1.057 quart ....qt

PRESSURE
1 kilopascal = .145/in²
  (1 bar = 101.325 kilopascals)

STRESS
1 megapascal or
1 newton/millimeter² = 145 pound/in² (psi) ....psi
  (1 N/mm² = 1 MPa)

POWER
1 kilowatt = 1.341 horsepower (550 ftlb/s) ..hp
  1 watt = 1 Nm/s

ENERGY (WORK)
1 joule = 0.000 947 8 British Thermal Unit.BTU
  (1 J = 1 W s)

FORCE
1 newton = 0.2248 pounds force ...... lb force

TORQUE OR BENDING MOMENT
1 newton meter = 0.7376 pound-foot ....(lb-ft)

TEMPERATURE
$t_C = (t_F - 32)/1.8$

English to Metric

LENGTH
1 inch = 25.4 millimeters ....mm
1 foot = 0.3048 meters ....m
1 yard = .9144 meters ....m
1 mile = 1.608 kilometers ....km

AREA
1 foot² = 0.0929 meter² ....m²
1 acre = 0.4047 hectare ....ha
  (hectare = 10,000 m²)

MASS WEIGHT
1 pound = 0.4535 kilograms ....kg
1 ton (2000 lb) = 0.9071 tonnes ....t

VOLUME
1 foot³ = 0.028 32 meter³ ....m³
1 yard³ = 0.7646 meter³ ....m³
1 bushel = 0.035 24 meter³ ....m³
1 bushel = 35.24 liter L
1 quart = 0.9464 liter L
1 gallon = 3.785 liters L

PRESSURE
1 pound/inch² = 6.895 kilopascals
          = 0.06895 bars

STRESS
1 pound/in² (psi) = 0.006 895 megapascal ....MPa
              or newton/mm² ....N/mm²
  (1 N/mm² = 1 MPa)

POWER
1 horsepower (550 ftlb/s) = .7457 kilowatt ..kW
  (1 watt = 1 Nm/s)

ENERGY (WORK)
1 British Thermal Unit = 1055 joules ....J
  (1 J = 1 W s)

FORCE
1 pound = 4.448 newtons ....N

TORQUE OR BENDING MOMENT
1 pound-foot = 1.356 newton-meters ....Nm

TEMPERATURE
$t_F = 1.8\ t_C + 32$

UNIFIED INCH BOLT AND CAP SCREW TORQUE VALUES

SAE Grade and Head Markings	NO MARK	1 or 2[b]	5 5.1 5.2	8 8.2
SAE Grade and Nut Markings	NO MARK	2	5	8

Size	Grade 1				Grade 2[b]				Grade 5, 5.1, or 5.2				Grade 8 or 8.2			
	Lubricated[a]		Dry[a]		Lubricated[a]		Dry[a]		Lubricated[a]		Dry[a]		Lubricated[a]		Dry[a]	
	N·m	lb-ft	N·m	lb-ft	N·m	lb-ft	N·m	lb-ft	N·m	lb-ft	N·m	lb-ft	N·m	lb-ft	N·m	lb-ft
1/4	3.7	2.8	4.7	3.5	6	4.5	7.5	5.5	9.5	7	12	9	13.5	10	17	12.5
5/16	7.7	5.5	10	7	12	9	15	11	20	15	25	18	28	21	35	26
3/8	14	10	17	13	22	16	27	20	35	26	44	33	50	36	63	46
7/16	22	16	28	20	35	26	44	32	55	41	70	52	80	58	100	75
1/2	33	25	42	31	53	39	67	50	85	63	110	80	120	90	150	115
9/16	48	36	60	45	75	56	95	70	125	90	155	115	175	130	225	160
5/8	67	50	85	62	105	78	135	100	170	125	215	160	215	160	300	225
3/4	120	87	150	110	190	140	240	175	300	225	375	280	425	310	550	400
7/8	190	140	240	175	190	140	240	175	490	360	625	450	700	500	875	650
1	290	210	360	270	290	210	360	270	725	540	925	675	1050	750	1300	975
1-1/8	470	300	510	375	470	300	510	375	900	675	1150	850	1450	1075	1850	1350
1-1/4	570	425	725	530	570	425	725	530	1300	950	1650	1200	2050	1500	2600	1950
1-3/8	750	550	950	700	750	550	950	700	1700	1250	2150	1550	2700	2000	3400	2550
1-1/2	1000	725	1250	925	990	725	1250	930	2250	1650	2850	2100	3600	2650	4550	3350

DO NOT use these values if a different torque value or tightening procedure is listed for a specific application. The torque values listed are for general use only. Check the tightness of the cap screws periodically.

Shear bolts are designed to fail under predetermined loads. Always replace the shear bolts with an identical grade.

Fasteners should be replaced with the same or higher grade. If higher grade fasteners are used, these should only be tightened to the strength of the original.

Make sure the fastener threads are clean and you properly start thread engagement. This will prevent them from failing when tightening.

Tighten the plastic insert or crimped steel-type lock nuts to approximately 50 percent of amount shown in chart. Tighten the toothed or serrated-type lock nuts to full torque value.

[a] "Lubricated" means coated with a lubricant such as engine oil, or fasteners with phosphate and oil coatings. "Dry" means plain or zinc plated without any lubrication.

[b] Grade 2 applies for hex cap screws (not hex bolts) up to 152 mm (6-in.) long. Grade 1 applies for hex cap screws over 152 mm (6-in.) long, and for all other types of bolts and screws of any length.

METRIC BOLT AND CAP SCREW TORQUE VALUES

Property Class and Head Markings	4.8	8.8	9.8	10.9	12.9
Property Class and Nut Markings	5	10		10	12

Size	Class 4.8				Class 8.8 or 9.8				Class 10.9				Class 12.9			
	Lubricated[a]		Dry[a]		Lubricated[a]		Dry[a]		Lubricated[a]		Dry[a]		Lubricated[a]		Dry[a]	
	N·m	lb-ft	N·m	lb-ft	N·m	lb-ft	N·m	lb-ft	N·m	lb-ft	N·m	lb-ft	N·m	lb-ft	N·m	lb-ft
M6	4.8	3.5	6	4.5	9	6.5	11	8.5	13	9.5	17	12	15	11.5	19	14.5
M8	12	8.5	15	11	22	16	28	20	32	24	40	30	37	28	47	35
M10	23	17	29	21	43	32	55	40	63	47	80	60	75	55	95	70
M12	40	29	50	37	75	55	95	70	110	80	140	105	130	95	165	120
M14	63	47	80	60	120	88	150	110	175	130	225	165	205	150	260	190
M16	100	73	125	92	190	140	240	175	275	200	350	225	320	240	400	300
M18	135	100	175	125	260	195	330	250	375	275	475	350	440	325	560	410
M20	190	140	240	180	375	275	475	350	530	400	675	500	625	460	800	580
M22	260	190	330	250	510	375	650	475	725	540	925	675	850	625	1075	800
M24	330	250	425	310	650	475	825	600	925	675	1150	850	1075	800	1350	1000
M27	490	360	625	450	950	700	1200	875	1350	1000	1700	1250	1600	1150	2000	1500
M30	675	490	850	625	1300	950	1650	1200	1850	1350	2300	1700	2150	1600	2700	2000
M33	900	675	1150	850	1750	1300	220	1650	2500	1850	3150	2350	2900	2150	3700	2750
M36	1150	850	1450	1075	2250	1650	2850	2100	3200	2350	4050	3000	3750	2750	4750	3500

 CAUTION: Use only metric tools on metric hardware. Other tools may not fit properly. They may slip and cause injury.

DO NOT use these values if a different torque value or tightening procedure is listed for a specific application. The torque values listed are for general use only. Check the tightness of the cap screws periodically.

Shear bolts are designed to fail under predetermined loads. Always replace the shear bolts with an identical grade.

Fasteners should be replaced with the same or higher grade. If higher grade fasteners are used, these should only be tightened to the strength of the original.

Make sure the fastener threads are clean and you properly start thread engagement. This will prevent them from failing when tightening.

Tighten the plastic insert or crimped steel-type lock nuts to approximately 50 percent of amount shown in chart. Tighten the toothed or serrated-type lock nuts to full torque value.

a *"Lubricated means coated with a lubricant such as engine oil, or fasteners with phosphate and oil coatings. "Dry means plain or zinc plated without any lubrication.*

INDEX

ANSWERS TO TEST YOURSELF QUESTIONS

CHAPTER 1

1. The four principles:

   a. Liquids have no shape of their own.
   b. Liquids will not compresss.
   c. Liquids transmit applied pressure in all directions.
   d. Liquids provide great increase in work force.

2. Force = Area x Pressure.

3. The four components:

   a. A *pump* to push the fluid through the system.
   b. A *cylinder* (or motor) to convert fluid movement into work.
   c. *Valves* to control fluid pressure and flow.
   d. A *reservoir* to store the fluid.

4. Answer to first blank: "open".
   Answer to second blank: "closed".

5. In an open-center system, the control valve is open during neutral and oil flows through and back to reservoir. In a closed-center system, the control valve is closed off during neutral and no oil is pumped through.

6. A fluid is normally *pushed* into a pump by gravity.

7. No. A pump creates flow. Pressure is caused by resistance to flow.

8. b. drop

CHAPTER 2

1. A hydraulic pump converts **mechanical** force into **hydraulic** force.

2. True.

3. False. They have no way of preventing back-feeding of oil.

4. Gear, vane and piston pumps are generally used.

5. An external gear pump has a gear on a gear and these gears rotate in the opposite direction of each other. The internal gear pump has a gear within a gear and these gears both rotate in the same direction.

6. Unbalanced — The eccentricity of the rotor to the rotor ring can be changed to vary the displacement.

7. In "axial" pumps, the pistons are parallel with the axis of their cylinder block. In "radial" pumps, the pistons are perpendicular to the axis of their cylinder block.

8. Gear

9. Human error is the No. 1 cause of pump failures. Specifically, improper condition of the hydraulic fluid is the most frequent single cause of pump failures.

10. It will decrease the life of the pump bearings and thus the pump to ⅛ its normal service life.

11. Reduce — ½

12. A pump usually cavitates because of a restricted inlet line. This allows air spaces to develop in the incoming fluid. These spaces collapse when reaching a higher pressure area, thus causing a jarring effect on the pump which may result in damage to the pump parts.

CHAPTER 3

I. Introduction

1. Pressure control valves, directional control valves, and volume control valves.

II. Pressure Control Valves

1. A. Limit system pressure, B. Reduce system pressure, C. Unload a pump, D. Set pressure for secondary circuit.

2. A. Direct acting, B. Pilot operated.

3. Cracking pressure is the pressure at which a valve first begins to pass a flow of oil. Full flow pressure is the pressure at which a valve passes its full rated capacity.

4. Both. Depending upon the application, a valve can be rated either way.

5. If a direct-acting relief valve fails, system pressure merely drops, resulting in loss of hydraulic functions. In other words, the pressure does not become dangerously high when a relief valve fails.

6. They are almost exact opposites. Relief valve — normally closed, reducing valve — normally open. Relief valve — sensing line from inlet, reducing valve — sensing line from outlet.

III. Directional Control Valves

1. A. Check valve (poppet type), B. Rotary type, C. Sliding spool type, D. Pilot-controlled poppet valve, E. Electro-hydraulic valve.

2. The simple check valve.

3. Open center and closed center.

CHAPTER 3 (cont.)

4. In neutral, pump flow passes through the open center valve, but is blocked in the closed center spool valve.

IV. Volume Control Valves

1. Restricting fluid flow or diverting it.

2. "Noncompensated" valves control flow simply by restricting it. "Compensated" valves control flow but also regulate it in response to pressure changes.

3. A needle valve.

4. By pressure drop through an orifice.

V. Microprocessor Controlled Valves

1. Elimination of mechanical linkage.

2. Flexibility in design.

3. Controller can be "taught."

4. Greater accuracy of control.

CHAPTER 4

1. False. Cylinders convert fluid power to mechanical power.

2. First blank — "straight." Second blank — "rotary."

3. First blank — "double." Second blank — "single."

4. One side is filled with hydraulic oil; the other side is filled with air.

5. The rod fills an area of the piston which is not exposed to pressure oil. So the stroke produced on this side is less powerful, but usually faster (since less oil is needed to move the piston).

6. It slows down the cylinder movement at the end of its stroke by partly closing off the discharged oil in front of the rotating vane.

CHAPTER 5

1. A hydraulic motor converts *fluid* force into *mechanical* force.

2. The motor works in reverse when compared to a pump. The pump draws in fluid and pushes it out, while the motor has fluid forced in but exhausts it out. In other words, the pump *drives* its fluid, while the motor is *driven by* its fluid.

3. Gear, vane, and piston types.

4. Torque is a measure of the turning force exerted at the outer edge of the motor drive shaft.

5. False. Though very similar, pumps often need

special bearings and other parts when used as motors. To avoid trouble, use pumps as pumps and motors as motors.

6. Fixed-mount and steerable.

CHAPTER 6

1. If hydraulic oil and oxygen mix under pressure, an explosion could take place. Instead, use an inert gas such as dry nitrogen.

2. When air is compressed, water vapor in the air condenses and can cause rust. This in turn may damage the seals and ruin the accumulator, and once air leaks into the oil, the oil will be oxidized.

3. Relieve all hydraulic pressure in the accumulator.

4. 1) Store energy, 2) Absorb shocks, 3) Build pressure gradually, 4) Maintain constant pressure.

CHAPTER 7

1. To keep the oil clean.

2. A full-flow system filters all the oil for each cycle. A bypass system filters only a small part of the oil.

3. It opens to allow oil to flow around the filter when the filter becomes plugged.

4. Surface filters stop dirt outside the filter. Depth filters stop dirt inside like a sponge.

5. Because they chemically remove desirable additives from the oil.

6. A micron (0.00004 inch).

7. A piece of dirt will wear away moving parts in the hydraulic system. Each new wear particle will in turn cause more wear particles and the contamination will continue to multiply.

8. Cap or plug the lines and connectors.

CHAPTER 8

1. It helps to keep oil clean and dissipates heat and air from the oil.

2. Air and water.

3. 1. Inner tube, 2. Reinforcement, 3. Outer cover.

4. 1. Avoid taut hose, 2. Avoid loops, 3. Avoid twisting, 4. Avoid rubbing, 5. Avoid heat, 6. Avoid sharp bends.

5. False. The zinc coating may flake or scale into the oil and damage precision hydraulic parts.

CHAPTER 8 (cont.)

6. True. For ease of removal, tubing should have an angle or two. This also allows for expansion and contraction.

CHAPTER 9

1. First blank — "Static." Second blank — "dynamic."

2. O-ring.

3. True. A small amount of leakage forms a film of oil which lubricates the moving parts. However, oil drips are not permissible.

4. False. All seals that are *disturbed* during a repair job should be replaced, whether they appear damaged or not.

CHAPTER 10

1. Highly refined petroleum oil plus additives.

2. Thinner.

3. Thicker.

4. Harmful contaminants develop.

5. Each product has specific requirements.

CHAPTER 11

1. 1) Not enough oil in the reservoir, 2) Clogged or dirty filters, 3) Loose intake lines, 4) Incorrect oil in the system.

2. False. Most solvents and cleaners are NOT recommended because they are poor lubricants and may damage seals or cause rusting of parts and breakdown of the oil. Instead, use hydraulic oil as a flushing agent.

3. Heat.

4. a — 3. b — 1. c — 2.

CHAPTER 12

1. 1) Know the system, 2) Ask the operator, 3) Operate the machine, 4) Inspect the machine, 5) List the possible causes, 6) Reach a conclusion, 7) Test your conclusion.

2. During *none* of these steps. Do all these things *before* you start repairing the system.

3. False. Test the *pump* first. It is the generating force for the whole system and affects the operation of all other system components.

4. First blank — "system". Second blank — "circuit".

5. A systematic quick-check where you use your senses only.

6. Look, Listen and Feel.

CHAPTER 13

1. Internationally recognized; cross language barriers; simplify design, fabrication, analysis, and function of circuits; show connections, flow paths, and functions of components

2. By the placement of the arrowheads inside the circle. Arrowheads pointing out mean that it is a pump. Arrowheads pointing in to the center of the circle mean that it is a motor.

3. Indicates that the pump or motor is a variable displacement type

4. False. Four squares represent a four-position valve.

5. Three — one for each position

6. Only one because the valve can only be in one valving position at a time

7. Nonactuated

8. True

9. Hydraulic oil reservoir or tank

10. American National Standards Institute

CHAPTER 14

1. Be careful. Safety hazard.

2. Caution — Warning — Danger.

3. A piece of cardboard.

4. Energy.

5. A place where two objects move toward each other, or where one object moves toward a stationary object.

6. a. Lower all equipment to the ground.

   b. Stop engine and remove key.

   c. Disconnect the battery ground strap.

   d. Hang a "Do Not Operate" sign, or tag in the operator station.

SAFETY
live with it